THE ENVIRONMENTAL CRISIS

Other Titles in the Greenwood Press Guides
to Historic Events of the Twentieth Century
Randall M. Miller, Series Editor

The Civil Rights Movement
Peter B. Levy

The Holocaust
Jack R. Fischel

The Breakup of Yugoslavia and the War in Bosnia
Carole Rogel

Islamic Fundamentalism
Lawrence Davidson

Frontiers of Space Exploration
Roger D. Launius

The Collapse of Communism in the Soviet Union
William E. Watson

Origins and Development of the Arab-Israeli Conflict
Ann M. Lesch and Dan Tschirgi

The Rise of Fascism in Europe
George P. Blum

The Cold War
Katherine A.S. Sibley

The War in Vietnam
Anthony O. Edmonds

World War II
Loyd E. Lee

The Unification of Germany, 1989–1990
Richard A. Leiby

THE ENVIRONMENTAL CRISIS

Miguel A. Santos

Greenwood Press Guides to
Historic Events of the Twentieth Century
Randall M. Miller, Series Editor

Greenwood Press
Westport, Connecticut • London

Library of Congress Cataloging-in-Publication Data

Santos, Miguel A.
    The environmental crisis / Miguel A. Santos.
        p.   cm.—(Greenwood Press guides to historic events of the
    twentieth century, ISSN 1092–177X)
    Includes bibliographical references and index.
    ISBN 0–313–30151–4 (alk. paper)
    1. Environmental policy.   2. Environmental policy—United States.
    3. Environmentalism.   4. Environmentalists—Biography.   I. Title.
    II. Series
    GE170.S367 1999
    363.7—dc21   98–28768

British Library Cataloguing in Publication Data is available.

Library of Congress Catalog Card Number: 98–28768
ISBN: 0–313–30151–4
ISSN: 1092–177X

First published in 1999

Greenwood Press, 88 Post Road West, Westport, CT 06881
An imprint of Greenwood Publishing Group, Inc.

Printed in the United States of America

The paper used in this book complies with the
Permanent Paper Standard issued by the National
Information Standards Organization (Z39.48–1984).

10 9 8 7 6 5 4 3 2 1

ADVISORY BOARD

Michael Adas
Professor of History, Rutgers University

Ross E. Dunn
Professor of History, San Diego State University

Howard Spodek
Professor of History and Urban Studies, Temple University

Contents

Series Foreword

As the twenty-first century approaches, it is time to take stock of the political, social, economic, intellectual, and cultural forces and factors that have made the twentieth century the most dramatic period of change in history. To that end, the Greenwood Press Guides to Historic Events of the Twentieth Century presents interpretive histories of the most significant events of the century. Each book in the series combines narrative history and analysis with primary documents and biographical sketches, with an eye to providing both a reference guide to the principal persons, ideas, and experiences defining each historic event, and a reliable, readable overview of that event. Each book further provides analyses and discussions, grounded in both primary and secondary sources, of the causes and consequences, in thought and action, that give meaning to the historic event under review. By assuming a historical perspective, drawing on the latest and best writing on each subject, and offering fresh insights, each book promises to explain how and why a particular event defined the twentieth century. No consensus about the meaning of the twentieth century emerges from the series, but, collectively, the books identify the most salient concerns of the century. In so doing, the series reminds us of the many ways those historic events continue to affect our lives.

Each book follows a similar format designed to encourage readers to consult it both as a reference and a history in its own right. Each volume opens with a chronology of the historic event, followed by a narrative overview, which also serves to introduce and examine briefly the main themes and issues related to that event. The next set of chapters is composed of topical essays, each analyzing closely an issue or problem of interpretation introduced in the opening chapter. A concluding chapter suggesting the long-term implications and meanings of the historic event brings the strands of the preceding chapters together while placing the event in the larger historical context. Each book also includes a section of short biographies of the principal persons related to the event, followed by a section introducing and reprinting key historical documents illustrative of and pertinent to the event. A glossary of selected terms adds to the utility of each book. An annotated bibliography—of significant books, films, and CD-ROMs—and an index conclude each volume.

The editors made no attempt to impose any theoretical model or historical perspective on the individual authors. Rather, in developing the series, an advisory board of noted historians and informed high school history teachers and public and school librarians identified the topics needful of exploration and the scholars eminently qualified to examine those events with intelligence and sensitivity. The common commitment throughout the series is to provide accurate, informative, and readable books, free of jargon and up to date in evidence and analysis.

Each book stands as a complete historical analysis and reference guide to a particular historic event. Each book also has many uses, from understanding contemporary perspectives on critical historical issues, to providing biographical treatments of key figures related to each event, to offering excerpts and complete texts of essential documents about the event, to suggesting and describing books and media materials for further study and presentation of the event, and more. The combination of historical narrative and individual topical chapters addressing significant issues and problems encourages students and teachers to approach each historic event from multiple perspectives and with a critical eye. The arrangement and content of each book thus invite students and teachers, through classroom discussions and position papers, to debate the character and significance of great historic events and to discover for themselves how and why history matters.

The series emphasizes the main currents that have shaped the modern world. Much of that focus necessarily looks at the West, especially Europe and the United States. The political, commercial, and cultural expansion of the West wrought largely, though not wholly, the most fundamental changes of the century. Taken together, however, books in the series reveal the interactions between Western and non-Western peoples and society, and also the tensions between modern and traditional cultures. They also point to the ways in which non-Western peoples have adapted Western ideas and technology and, in turn, influenced Western life and thought. Several books examine such increasingly powerful global forces as the rise of Islamic fundamentalism, the emergence of modern Japan, the Communist revolution in China, and the collapse of communism in eastern Europe and the former Soviet Union. American interests and experiences receive special attention in the series, not only in deference to the primary readership of the books but also in recognition that the United States emerged as the dominant political, economic, social, and cultural force during the twentieth century. By looking at the century through the lens of American events and experiences, it is possible to see why the age has come to be known as "The American Century."

Assessing the history of the twentieth century is a formidable prospect. It has been a period of remarkable transformation. The world broadened and narrowed at the same time. Frontiers shifted from the interiors of Africa and Latin America to the moon and beyond; communication spread from mass circulation newspapers and magazines to radio, television, and now the Internet; skyscrapers reached upward and suburbs stretched outward; energy switched from steam, to electric, to atomic power. Many changes did not lead to a complete abandonment of established patterns and practices so much as a synthesis of old and new, as, for example, the increased use of (even reliance on) the telephone in the age of the computer. The automobile and the truck, the airplane, and telecommunications closed distances, and people in unprecedented numbers migrated from rural to urban, industrial, and ever more ethnically diverse areas. Tractors and chemical fertilizers made it possible for fewer people to grow more, but the environmental and demographic costs of an exploding global population threatened to outstrip natural resources and human innovation. Disparities in wealth increased, with developed nations prospering and underdeveloped nations starving. Amid the crumbling of former European colonial empires, Western technology, goods, and culture increasingly enveloped the globe, seeping into, and undermining, non-Western cultures—a process that contributed to a surge of religious fundamentalism and ethno-nationalism in the Middle East, Asia, and Africa. As people became more alike, they also became more aware of their differences. Ethnic and religious rivalries grew in intensity everywhere as the century closed.

The political changes during the twentieth century have been no less profound than the social, economic, and cultural ones. Many of the books in the series focus on political events, broadly defined, but no books are confined to politics alone. Political ideas and events have social effects, just as they spring from a complex interplay of non-political forces in culture, society, and economy. Thus, for example, the modern civil rights and woman's rights movements were at once social and political events in cause and consequence. Likewise, the Cold War created the geopolitical framework for dealing with competing ideologies and nations abroad and served as the touchstone for political and cultural identities at home. The books treating political events do so within their social, cultural, and economic contexts.

Several books in the series examine particular wars in depth. Wars are defining moments for people and eras. During the twentieth century war became more widespread and terrible than ever before, encouraging new efforts to end war through strategies and organizations of international cooperation and disarmament while also fueling new ideologies and instruments of mass persuasion that fostered distrust and festered old national rivalries. Two world wars during the century redrew the political map, slaughtered or uprooted two generations of people, and introduced and hastened the development of new technologies and weapons of mass destruction. The First World War spelled the end of the old European order and spurred communist revolution in Russia and fascism in Italy, Germany, and elsewhere. The Second World War killed fascism and inspired the final push for freedom from European colonial rule in Asia and Africa. It also led to the Cold War that suffocated much of the world for almost half a century. Large wars begat small ones, and brutal totalitarian regimes cropped up across the globe. After (and in some ways because of) the fall of communism in eastern Europe and the former Soviet Union, wars of competing cultures, national interests, and political systems persisted in the struggle to make a new world order. Continuing, too, has been the belief that military technology can achieve political ends, whether in the superior American firepower that failed to "win" in Vietnam or in the American "smart bombs" and other military wizardry that "won" in the Persian Gulf.

Another theme evident in the series is that throughout the century nationalism has continued to drive events. Whether in the Balkans in 1914 triggering World War I or in the Balkans in the 1990s threatening the post–Cold War peace—or in many other places—nationalist ambitions and forces would not die. The persistence of nationalism is yet another reminder of the many ways that the past becomes prologue.

We thus offer the series as a modern guide to and interpretation of the historic events of the twentieth century and as an invitation to consider how and why those events have defined not only the past and present but also charted the political, social, intellectual, cultural, and economic routes into the next century.

Randall M. Miller
Saint Joseph's University, Philadelphia

Preface

The twentieth century has been the period in which the terms "ecology" and "environment" have become household words as well as potent political forces. During the century, the United States issued the first major federal response that called for a national environmental policy, the National Environmental Policy Act (NEPA) of 1969. With the enactment of NEPA, the United States began a "Magna Carta–like" federal program on environmental protection and management. In 1992 the biggest and most important environmental conference, the U.N. Conference on Environment and Development, also known as the "Earth Summit," brought together six thousand delegates from over 170 nations to discuss and negotiate such topics as climate change, biodiversity, deforestation, and economic relations between developed and less-developed nations.

During the late twentieth century, a consensus seemed to emerge throughout most of the United States and the world that it was time to combat pollution and seek a balance with nature. In hindsight, it is not clear what led to this unexpected change in humanity's worldview. It is likely that forces were multiple and complex, including the academic credibility ecology gained from professional scientists of the time as well as social, economic, and political factors.

Of all the "great events" of the twentieth century, perhaps the most significant to the future of this planet is the way in which humankind has tried to achieve equilibrium with nature. This book covers many of the fundamental concerns that have emerged regarding the relationship between society, natural resources, and pollution. It examines the main causes of environmental concern and the key players who brought about the surge of environmental consciousness in the twentieth century. This book also considers the new multidisciplinary ecological ethic and outlook that eventually brought about major modifications in both national and international laws and priorities. It also includes explicit discussions of scientific, social, economic, legal, ethical, and historical dimensions of the environmental crisis.

The forces that brought the environmental crisis to society's attention grew out of the vast social and economic changes that took place worldwide. From an ecological perspective, they can be divided into three fundamental categories: the concerns for the vanishing wilderness, pollution, and overpopulation. Interwoven throughout these three categories is a fourth less tangible concern, involving long-term coexistence between humanity and its environment. This issue will be studied under the concept of self-sustainable systems.

This volume takes its structure from these four fundamental concerns and looks at the principal actors of the environmental vanguard in the twentieth century. Chapter 1 gives a historical overview of the environmental crisis, concentrating on the chronological development of environmental policies in the United States and elsewhere. It summarizes the role that key environmentalists and environmental organizations have played in the development of environmental policies during the twentieth century.

Chapter 2 describes the first surge in society's environmental consciousness—namely, its concern for the vanishing wilderness. It also includes a discussion of conditions and trends in biodiversity and the origins of the concerns for the policies and programs that protect biodiversity.

Chapter 3 describes how pollutants emitted by faulty technology galvanized the environmentalists and spurred a second wave of environmental legislation and treaties. Its objective is to explain several events that heralded this new awareness of pollutants. These events include the 1962 publication of *Silent Spring* by Rachel Carson, the international concern for the depletion of the ozone layer, and the issue of global climate change.

Chapter 4 deals with the overpopulation problem and includes a discussion of the controversial population policies of various nations, including India, China, and the United States. Its major focus is to explore the implications of population ecology principles of human ecology. This chapter also reviews the growth patterns of various nations.

Chapter 5 constructs an analytical framework of a self-sustainable human system. The chapter also presents an overview of scientific and socioeconomic dimensions for building a sustainable human-environmental system. It summarizes many of the changes that are necessary to create a sustainable society that meets its needs without degrading the environment for future generations.

This book also includes several special features. A chapter of biographies contains brief biographical sketches of the key people involved in the development of environmentalism in the twentieth century. The documents section includes excerpts from the most important environmental laws of the United States, treaties, and environmental declarations. The glossary defines key terms. Finally, an annotated bibliography directs the reader to significant, and widely accessible, print and nonprint sources on the environmental crisis.

I am grateful to a number of people for their varied assistance in this project. Professor Randall Miller of Saint Joseph's University provided considerable editing of the earlier drafts of this manuscript. I acknowledge the important contributions of students in my ecology and environmental studies courses over the last couple of years whose responses have had a great influence on the writing of this book. Finally, I owe the greatest debt of gratitude to my wife, Zulema, for her day-to-day encouragement. I appreciate the help of our two daughters, Brenda and Cynthia, for the preparation of the manuscript.

Chronology of Events

A.D.1	World population at 200 million.
1750	World population at 791 million.
1798	Thomas Malthus publishes *Essay on the Principle of Population*.
1800	World population at 978 million.
1845	Irish potato famine.
1854	Henry David Thoreau publishes *Walden*.
1859	Charles Darwin publishes *On the Origin of Species*.
1864	George Perkins Marsh publishes *Man and Nature*.
1869	Ernst Haeckel first uses the term "ecology" in an address.
1870	The first official wildlife sanctuary is established at Lake Merritt (California).
1872	Yellowstone, the world's first national park, is established (largely in northwestern Wyoming).
1892	The oldest and largest nongovernmental environmental organization, Sierra Club, is created.
1900	The Lacey Act, a landmark in federal regulation, is passed.
1900	World Population at 1.650 billion.
1903	The first federal wildlife refuge, Pelican Island, Florida, is established by President Theodore Roosevelt.

1915   The Ecological Society of America is formed, with Victor Shel-
        ford elected as the first president.

1916   The U.S. National Park Service is established.

1918   The Migratory Bird Treaty protecting nongame birds that migrate
        between the United States and Canada is signed.

1929   Vladimir Vernadsky publishes *La Biosphere*.

1933   Civilian Conservation Corps and Soil Service (later renamed the
        Soil Conservation Service) are created.

1949   Aldo Leopold's *Sand County Almanac* is published posthumously.

1953   Eugene Odum publishes *Fundamentals of Ecology*.

1962   Rachel Carson publishes *Silent Spring*.

1964   The Wilderness Act is enacted.

1966   Species Conservation Act is enacted. (In 1973 it is amended and
        renamed Endangered Species Act.)

1967   *Torrey Canyon* oil spill (36 million gallons) on the British and
        French seashores.

1968   Paul Ehrlich publishes *The Population Bomb*.

1968   Biosphere Conference in Paris, sponsored by United Nations Edu-
        cational, Scientific, and Cultural Organization (UNESCO).

1968   Garrett Hardin publishes "The Tragedy of the Commons."

1969   Oil spill in Santa Barbara, California (amount undetermined).

1969   Oil-slicked and debris-choked section of the Cuyahoga River near
        Cleveland bursts into flames.

1970   U.S. Environmental Protection Agency is established.

1970   The National Environmental Policy Act becomes law.

1970   First Earth Day is celebrated.

1971   *The Closing Circle* by Barry Commoner is published.

1972   *The Limits to Growth* by a group of scientists from the Massachu-
        setts Institute of Technology appears.

1972   New Zealand forms the first national environmental party, called
        Values.

1972   United Nations Stockholm Conference on the Human Environ-
        ment.

1972	United Nations Environment Programme (UNEP) is established by the General Assembly.
1972	First returnable bottle legislation (Oregon).
1973	Organization of Petroleum Exporting Countries (OPEC) engineers an "energy crisis."
1977	North Sea oil spill (8.2 million gallons).
1978	Toxic dump tragedy at Love Canal in Buffalo, New York.
1978	Nuclear accident at Three Mile Island near Harrisburg, Pennsylvania.
1978	*Amoco Cadiz* oil spill off France (70 million gallons).
1984	Union Carbide pesticide plant disaster at Bhopal, India.
1984	Famine in the Sahel, Africa.
1984	First report on the *State of the World* by the Worldwatch Institute.
1985	Discovery of an "ozone hole" in the stratosphere over Antarctica.
1986	Nuclear power plant disaster at Chernobyl in the Ukraine.
1987	*Our Common Future* is published by the World Commission on Environment and Development.
1987	Montreal Protocol Treaty.
1988	International Panel on Climate Change is established by the World Meteorological Organization and the United Nations Environment Programme.
1989	*Exxon Valdez* oil spill, Alaska (11 million gallons).
1992	"Earth Summit" in Rio de Janeiro, Brazil.
1995	The first United Nations Conference of the Parties to the Framework Convention on Climate Change.
1995	The first Nobel Prize is awarded to environment-related research to Sherwood Roland, Mario Molina, and Paul Crutzen.
1997	Kyoto Protocol Treaty.
1998	World population at 6.0 billion.

THE ENVIRONMENTAL CRISIS EXPLAINED

I

Historical Overview of the Environmental Crisis

Society has had a detrimental effect upon the Earth through its exploiting or otherwise affecting most of its terrestrial and aquatic ecosystems. Whether negligently or intentionally, human activities have caused the deterioration of the natural environment and the extinction of many species. Pollutants may harm the senses of sight, taste, and smell and may also cause health hazards. The potential for climate change and resource depletion may eventually alter the fundamental framework of society as we know it. These potential effects will most likely become exacerbated by the end of the twenty-first century, when the Earth is expected to have a human biomass of thirty billion people.

It is natural to feel overwhelmed by the intricacies and complexities of the many environmental problems confronting society and by the belief that an individual is at the mercy of all the variables that threaten our society. Notwithstanding the fact that one individual might be helpless, virtually hundreds of thousands of individuals and organizations are attempting to resolve these problems and are achieving some measure of success along many lines. Many key players have emerged to deal with the numerous dimensions of the environmental problem. Some organizations are global in scope; others are national or local. Some have influence on the general public, some affect the government, and others act directly (for example, by buying up wilderness areas to protect them from development).

Therefore, environmental policy is strongly influenced by the interaction of many different people in private and public roles, although it is ultimately formed by public officials. The concerns about environmental policy can be divided into stages, which are largely determined by scientific, economic, and other social forces.

The aim of this chapter is to provide a narrative chronological overview of the environmental crisis of the twentieth century, describing and analyzing the environmental movement and policies, as well as introducing the principal actors and issues of the environmental vanguard. The first part of this chapter considers the history and development of environmental policies in the United States, and the second part looks at global environmental policies. Later chapters offer detailed descriptions of the environmental crisis and the development of environmental policies of developed and developing nations.

HISTORY AND DEVELOPMENT OF ENVIRONMENTALISM IN THE UNITED STATES

In America environmentalism is not a new social movement. Its origins date back to the nineteenth century, when trappers, fishermen, and naturalists campaigned against the unrestrained exploitation of America's pristine environments. In 1872 the United States established the world's first national park (Yellowstone), and in 1891 forty million acres in the first federal reserve were set aside for protection against the wanton deforestation of the time. Concerns for conservation were heightened during the post–Civil War decades by the emergence of powerful corporations bent on exploiting nature to its limits.

Intellectuals and environmentally minded thinkers such as George Perkins Marsh (1801–1882), Ralph Waldo Emerson (1803–1882), Henry David Thoreau (1817–1862), John Muir (1838–1914), and Gifford Pinchot (1865–1946) provided the philosophical framework of the nineteenth-century environmental movement, which stretched into the Progressive era before World War I. They were very influential in attracting public attention and affecting policymakers. Gifford Pinchot, for example, was both a forester and a politician; he was appointed Chief of the Division of Forestry (later renamed the U.S. Forest Service) by President Theodore Roosevelt. During Roosevelt's administration (1901–1909), the national forests were expanded to 190 million acres, which is their approximate size today (see Figure 1.1). These early conservation measures were guided primarily by the notion "that natural resources must be held in trust for future generations through careful use and management."[1]

Figure 1.1
National Forest System, 1891–1993

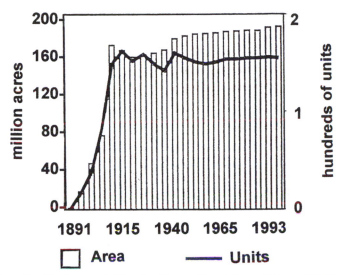

Source: Council on Environmental Quality. *Twenty-Fourth Annual Report*. Washington, DC:
U.S. Government Printing Office, 1993.

Another wave of environmentalism lasted from 1933 to 1945 when the
drought and dust storms of the 1930s pushed conservation issues into the
forefront. Federal New Deal programs responding to the Great Depression
used conservation policy to alleviate unemployment, justify stronger government controls over the economy, and integrate resource management.
During Franklin D. Roosevelt's administration, two billion trees were
planted by the Civilian Conservation Corps, a number of sewage treatment
programs were built by the Public Works Administration, and the national
forest was expanded.

Pollution concerns also have existed for a long time. Some large municipalities began treating their sewage in the nineteenth century, and the federal water pollution control statute was enacted in 1948. State and local
governments also made early attempts to control air pollution. Cincinnati
and Chicago enacted smoke control laws in 1881, and by 1912, twenty-
three of the twenty-eight U.S. cities with populations over two hundred
thousand had followed with similar ordinances. California's attempts to
combat smog began during the 1940s, and regulations on manufacturing
and motor vehicle control were added twenty years later. The first national
air pollution statute was enacted in the 1950s. The post–World War II eco-

nomic prosperity of the United States fueled this interest in environmental quality as more Americans spent leisure time outdoors and traveled the country and as higher levels of education led to a greater awareness of environmental issues.

Economic Prosperity and the Emergence of New Environmentalists

In spite of a few recessions, the United States underwent an era of extraordinary economic prosperity at the end of World War II. From 1950 to 1970, real income (personal disposable income per person, adjusted for inflation) increased by 52 percent, and average hours worked per week decreased by 7 percent. During the war, the majority of Americans had a hard time just paying their bills. After the war, the economy returned to domestic production, with increasing emphasis on consumer goods ranging from automobiles to televisions to toasters. A housing boom and the consumption of resources added to the prosperity. A relaxing of financial pressures and shortages after the war contributed to a dramatic increase in consumer spending, and Americans also had more leisure time.

Americans did many things in response to the new economic well-being and additional leisure time. First, there was the "baby boom" of 1947–1964, causing the United States population to increase from 145 million in 1947 to 192 million in 1964. Next, the growing middle-class population searched for nicer places to live outside the core city. Since the mid-1800s, commerce, industry, and finance, rather than agriculture, had slowly become the principal means to wealth and advancement. The stream of immigrants coming to the new America remained mostly in the cities where manufacturing jobs were to be found. Simultaneously, Americans left farms for the cities. The farm population gradually dropped from twenty-nine million (30 percent of the population) in 1900, to twenty-three million (15 percent of the population) in 1950.

As prosperity increased in the 1950s, many middle-class urban dwellers decided to spend a significant percent of their newly obtained disposable income on improving their quality of life, especially by realizing "the American Dream" of buying a home. Among all the available alternatives, many people chose to buy houses in the suburbs, where new home construction offered many amenities not available in older city housing and where loans were available from private and public lenders who preferred to underwrite "new construction." The highway building boom of the era, thanks to federal support, also made suburban living possible because it contributed to suburban sprawl and an automobile culture.

During the 1950s and 1960s, as the population grew by 19 percent, the number of people in the suburbs increased by 45 percent. The core of the city grew by 12 percent, while the nonmetropolitan population increased by only 6 percent. This suburban population also consumed more than its predecessors. Not only was money spent on buying appliances, televisions, and home improvements, but it was also spent on tourism and outdoor activities. One indication that Americans were interested in using their newly acquired economic power and free time to experience the outdoors was the number of people who were visiting national parks. Park visits surged sevenfold between 1955 and 1995 (see Figure 1.2).

The significant movement of the population to the suburbs, coupled with economic prosperity and the technological improvement that made it possible, began to take its toll. Suburban growth overwhelmed land resources. People who resided at the edge of cities observed cherished farmlands and wildlands vanishing before their eyes. Many of the earlier urban émigrés observed that the open spaces they had come to enjoy had disappeared. Those who moved to the suburbs later found it even more difficult to find a house within a reasonable commuting distance from the city. Traffic jams

Figure 1.2
Public Use of National Parks

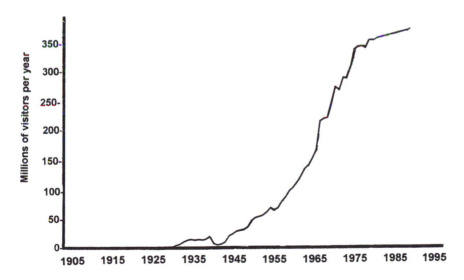

Source: Council on Environmental Quality. *Twenty-Fifth Anniversary Report*. Washington, DC: U.S. Government Printing Office, 1994.

became common, and new highways were built to improve traffic flow; but as soon as the new highways were built, that encouraged more suburban development. The new highways soon filled with traffic, along with all the noise and fumes.

All the while, many of the by-products of America's phenomenal growth also became apparent. From 1950 to 1970, as the population increased by 37 percent, the Gross National Product almost doubled. Everything in the economy grew: manufacturing, wholesale and retail trade, transportation, communication, utilities, construction, and fossil fuel consumption.

The environmental repercussions of prosperity and population growth and movement were becoming increasingly apparent. Charming landscapes and clear skies were obscured by smog. Health problems associated with high levels of air pollution surged. Polluted aquatic ecosystems were closed for fishing or bathing. Long stretches of unspoiled seashore became harder to find.

Chemical manufacturing experienced tremendous growth, with production rising five times between 1950 and 1970. After World War II, there was extensive manufacturing of synthetic organic materials. Chemists learned to engineer molecules "made to order" and began using their knowledge to convert raw materials into pesticides and fertilizers, as well as other substances that could be woven into clothes and wrap foods. These substances revolutionized American life. Today, some seventy thousand chemicals are in commercial use in the United States, and the number is growing.

As these substances became widely used, it became apparent that they would generate other concerns. One problem with synthetic organic compounds is their proper disposal. Plastic wrappers and bottles entering natural environments remain for years. In fact, a number of these items are nonbiodegradable and resist decomposition by microorganisms. There is also concern about the hazardous chemicals that some materials release into the environment. The unexpected side effects these substances pose for life are dangerous and sometimes unpredictable. There were reports of farm workers suffering from overexposure to pesticides and wildlife dying wherever pesticides were used.

During the late 1960s, many individuals who had previously sought the "American dream" with their extra money and leisure time found the quality of their environment slowly degrading. Yet, for most citizens, it was not precisely clear what the problem was and if it could be corrected.

Many books focused on the public's concern and even helped them understand some of the changes taking place. Four books of the period were well written and became classic environmental books, providing an intellectual framework for the environmental movement. Perhaps the most important was Rachel Carson's *Silent Spring* (1962), which vividly portrayed pollution as being not just an aesthetic problem, causing a bad smell or traffic jam, but also a threat to the complex processes of life itself. Her book awakened many Americans to the dangers of pesticides and inspired other writers to examine the environmental costs of unchecked growth and development. Paul Ehrlich's *The Population Bomb* (1968) stressed that environmental problems are primarily caused by overpopulation. He predicted that millions of people would starve in the 1970s due to a lack of resources necessary to feed them. Barry Commoner's *The Closing Circle* (1971) blamed the environmental dilemma on two faults of synthetic products, such as detergents and plastics. The first is that they require heavy dependence on artificial energy sources, such as fossil fuels (as opposed to the slower, less powerful energy capacities of nature such as photosynthesis). The second fault is that synthetic products do not biodegrade. Donnella Meadows, Dennis Meadows, Jorgen Randers, and William Behrens in *The Limits to Growth* (1972) used computer models to forecast population growth, resource depletion, food supply, capital investment, and pollution and described the coming limitations to conventional growth.

These books, benchmark events in themselves, had immense impact on the environmental movement. Written for general readers rather than for scholars, they were the first environmental books to sell millions of copies. The public, now alert to environmental concerns, demanded governmental action to protect the environment.

All the authors were scientists with a keen ability to interpret for society the way the world functions and the fortunes of humans as they live in nature. These scientists had the ability to analyze what was believed to be environmentally benign behavior from the past and to prescribe correct forms of behavior for the future. Scientifically generated information gave scientific legitimacy to their interpretations of how the world operates. Whether they were fair and impartial or not, these twentieth-century "philosophers of the Earth" formed the vanguard in providing an analytical framework for the environmental movement.

Besides the publication of these books, a variety of other factors helped to mobilize the environmental movement. The 1960s, for example, was a time of social upheaval. The era produced a generation of questioning, socially concerned, and politically active young people. Campus unrest swept the nation as the civil rights movement spread, and it peaked with demonstrations protesting the Vietnam War. The counterculture was alienated by bureaucracy, the Watergate scandal, materialism, and especially the Vietnam War. For the younger generation in America, whom the war affected most intimately, this was a time of profound disillusionment with America's military-business complex and other "establishment" institutions that promoted an impersonal, corporate culture. The generation of the 1960s sought new social values and lifestyles. Many young people, called "hippies," experimented with Eastern religions, drugs, and sex and tried to establish a self-sustainable "counterculture" lifestyle by doing so.

The new environmentalists, with their opposition to society's misuse of technology, made a significant philosophical connection between themselves and the peace and civil rights movements of the 1960s. The synergism between these movements created a climate in which people felt they had the conviction to take control of their lives and to make changes in America. At the same time, however, the association—in the public's mind—of environmentalism with the radical politics and counterculture values of "the left" slowed its general appeal.

A series of highly publicized environmental disasters gave environmentalists credibility for their warnings of imminent dangers and insensitive, irresponsible industrial polluters. The 1969 oil spill in Santa Barbara, which fouled many miles of an exceptionally beautiful beach and killed marine life by the thousands; the 1978 toxic dump tragedy at Love Canal, New York; and the 1979 nuclear accident at Three Mile Island, Pennsylvania—all galvanized public awareness about environmental issues and spurred a strong antipollution movement.

These and other environmental events seemed to have struck a public nerve and dramatized what many people saw as negligent (or, even worse, intentional) pollution of the environment. Environmental activists emphasized images of unbridled polluters, a technology out of control, and a *laissez-faire* attitude of a government that tolerated, even encouraged, the wholesale pollution of air, water, and land.

These outpourings of societal interests in ecology set things in motion for the celebration of the first Earth Day on April 22, 1970. The event was the idea of Senator Gaylord Nelson of Wisconsin, one of the few environmental advocates in Congress at the time. Nelson recruited Denis Hays, a Stanford graduate and campus antiwar activist, to serve as national coordinator. The first Earth Day was a populist-based celebration.

On that day, America witnessed an unprecedented demonstration by millions of people from different socioeconomic backgrounds. The concern about the state of environmental quality was expressed by people—from kindergartners to university people. Teach-ins, litter collections, and environmental rallies were held in communities throughout the nation. Indeed, these acts of growing public dissatisfaction would propel environmentalism as a potent political force. Over the next few decades, Earth Day's significance as the focal point for a new political coalition became apparent. Today, however, though the American public remains generally supportive of environmental measures, it often disagrees on specific policies of conservation, antipollution control, and limiting development.

As the environmental historian Samuel Hays pointed out, the *antipollution movement* of the 1960s differed dramatically from the *conservation movement* of the nineteenth century in that the former was an effort on the part of leaders in science, technology, and government to bring about more-efficient development of physical resources, whereas the latter was a product of a fundamental change in public values in the United States that stressed the quality of the human environment. According to Hays, the conservation-of-wildlife movement was a feature of the human chronicle of production that emphasized efficiency, whereas the antipollution movement was a part of the chronicle of consumption that stressed new features of the standard of living. Hays analyzed the social roots of environmental affairs and stressed the transformation of values that took place:

Public interest in environmental affairs . . . stems from a desire to improve personal, family, and community life. The desires are neither ephemeral nor erratic; they are evident in many nations, first in the advanced industrial and consumer societies and then in more recent years in those of middle and even earlier stages of development. They express human wants and needs as surely as do demands for better housing, more satisfying leisure and recreation, improved household furnishings, better health, and a greater sense of well-being. We customarily associate these with human "progress," which normally is accepted as a fundamental concern unnecessary to explain away in other terms. An interest in the environmental quality of life is to be understood simply as an integral part of the drives inherent in persistent human aspiration and achievement.[2]

Before the 1960s, traditional conservation groups, such as the Sierra Club and the National Audubon Society, were primarily concerned with the preservation of unique ecosystems. After the 1960s, as concern about pollution intensified, their interests in wildlife expanded to include this important issue. This shift also prompted the development of new environmental organizations with new and diverse environmental concerns (see Primary Documents, Table 1). In fact, one of these new organizations, Environmental Action, evolved from a student-oriented group that had originally coordinated the first Earth Day.

These new environmentalists, individuals or members of established environmental groups, began to encourage research, teaching, and lobbying on a variety of areas. Leaders from these national environmental organizations interacted with members and groups at the state level, thus playing a key role in shaping public policy. For example, in 1969 the Citizen's Crusade for Clean Water, a coalition of thirty-eight environmental groups, successfully lobbied an increase in congressional appropriations for waste-water treatment plants from $214 million to $800 million.

In recent years, environmental efforts have influenced the development of environmental policy, and whether done intentionally or otherwise, this "new vanguard for a new society" has made it possible for the federal government to expand its power over environmental regulation.

The Emergence of Federal Powers to Regulate the Environment

After the 1960s, in response to the political activism from citizens and environmental groups, the U.S. Congress enacted and/or amended numerous laws that touched virtually every aspect of public responsibility for the protection and management of the environment—from clean air, to the disposal of solid wastes, to saving endangered species (see Primary Documents 17–24). The responsibility for continuous and systematic administration of these laws was largely delegated to public agencies created for just that purpose, such as the Environmental Protection Agency (see Primary Documents, Table 1).

The new environmental laws were unlike previous ones in many respects. First, they came in succession; second, most of them had effective means of enforcement; and third, the federal government became more involved in environmental protection. Pressure to involve the federal government in environmental quality came from the awareness that some forms of pollution were complex and required the type of research base that only the federal government could provide. Moreover, it became apparent to lawmakers that a coordinated system of pollution control was necessary.

For most of American history, local and state governments had assumed responsibility for environmental protection. Except in the management of the vast federal lands of the trans-Mississippi West and the protection of American fishing and coastal interests, the federal government's role in direct environmental control was limited. The federal government through such agencies as the U.S. Geological Survey sponsored scientific expeditions that mapped and identified the nation's huge land and natural resources, but most governmental interest in such work was to pave the way for settlement or use. Policies of conservation did not come until late in the nineteenth century.

By the time of the New Deal, the balance of public responsibility shifted to the federal government. Although the state and local governments remain important in the fight against industrial pollution, the nation in the last few decades has looked increasingly to the federal government for guidance. Since the 1960s, that guidance was expressed in a number of far-reaching environmental laws, and the creation of two public agencies.

In 1970 the Council on Environmental Quality was established to assume responsibility for advising, coordinating, and monitoring the federal agencies' implementation of the National Environmental Policy Act of 1969. During the same year, fifteen scattered executive departments and independent agencies were combined into a centralized federal entity—the Environmental Protection Agency (EPA). The agency assumed responsibility as a central independent agency for carrying out federal laws to protect the environment.

Most federal regulations that affect the environment are based on the Commerce Clause in the Constitution, which provides Congress with the power "to regulate commerce with foreign nations, and among the several states." Since the 1930s, U.S. Supreme Court decisions have greatly broadened that regulatory power. Today, the test is whether or not the activity sought to be regulated by Congress has some appreciable effect, either direct or indirect, on interstate commerce. Viewed under these new regulatory powers, many aspects of environmental law have commercial connotations. The national government may regulate almost any natural resource or pollutant affecting interstate commerce. The Clean Water Acts and Clean Air Acts are examples of their involvement under the Commerce Clause.

State authority to regulate commerce resides in the powers reserved to it by the Constitution. Thus, unlike the federal government, there do not have to be affirmative constitutional sanctions for the state's police powers. *Police powers* refers to a broad state power to regulate public health, safety, morals, and welfare. The states have, in turn, delegated a portion of their authority to local government units.

The Commerce Clause, as interpreted by the courts, gives Congress exclusive jurisdiction over foreign commerce. States and their local government units sometimes attempt to regulate foreign commerce. To illustrate, a nonfederal government may attempt to impose a tax on foreign natural resources that compete with those locally produced or mined. Such actions violate both the Commerce Clause and the Supremacy Clause of the Constitution, which invalidates any conflicting state law by virtue of preemption, meaning that federal law takes precedence over state law.

The effect of the Commerce Clause can be illustrated by looking at the U.S. Supreme Court's decision in *City of Philadelphia v. New Jersey* (1978). This case involved a New Jersey law prohibiting the import of most solid or liquid waste into the state. The statute was written in response to the use of New Jersey's landfills for the disposal of waste from the neighboring states of Pennsylvania and New York. Several users of the landfill sites, from New Jersey and outside the state, sued to have the legislation declared unconstitutional as a violation of the Commerce Clause.

The Supreme Court invalidated the legislation on the grounds that it discriminated against interstate commerce. The opinion, by Justice Potter Stewart, concluded:

The New Jersey law at issue in this case falls squarely within the area that the Commerce Clause puts off limits to state regulation. On its face, it imposes on out-of-state commercial interests the full burden of conserving the State's remaining landfill space. . . . The State has overtly moved to slow or freeze the flow of commerce for protectionist reasons.[3]

Though the Commerce Clause accords the national government's pervasive powers, it is not exclusive. In general, states may regulate interstate commerce when the commerce being regulated does not require uniform treatment throughout the United States, state regulation is not an undue burden on commerce, and Congress has not prohibited, either expressly or implicitly, the area of regulation.

The following illustrates how the Commerce Clause is being used to develop a federal-state partnership wherein each jurisdiction would be delegated regulatory duties that it was most competent to fulfill. In both the Clean Water and Clean Air Acts, the EPA sets the environmental standards. Congress envisioned that states were in a better position to understand regional water and air qualities, and their economic priorities than the federal government. Consequently, each state was given the task of developing management programs and has the primary responsibility to enforce the standards. States can set standards more stringent than the federal government, and they can regulate pollutants that are not regulated by the federal government. However, if states fail to establish inspection and control programs, the EPA can establish and enforce such programs.

The Noise Control Act of 1972 presented a different situation. Under this Act, the primary duty of noise control rests with state and local governments; however, the federal government can take action where it finds that national uniformity is required because the product will be used between states and it is a major noise source. To illustrate, because it is not feasible for manufacturers of trucks, buses, lawn mowers, bulldozers, and other heavy machinery to comply with fifty different state laws regulating noise pollution, the federal government steps in and regulates noise pollution emanating from these products. Unlike the Clean Water and Air Acts, the Noise Pollution Control Act forbids the state from developing standards that are stricter than the federal standard.

The Constitution also provides that "Congress shall have the power to lay and collect taxes." This taxing power also has been used to regulate the environment. To illustrate, high import duties can be used to decrease the importation of certain foreign goods that might be damaging to the environment. The internal revenue code is used to regulate economic activity that is detrimental to the environment. For example, the government gives business tax credits for pollution control equipment and solar heating devices and therefore encourages businesses to make certain kinds of investments.

When the federal or a state government exercises one of its legitimate powers (e.g., the taxing powers), the legislation must be reasonably, or at least plausibly, related to the raising of revenue. The legislation may have more than one objective, such as energy conservation or pollution control. But in order to pass the due process test, there must be some arguable rationale for explaining the tax as a revenue source.

After stating the taxing power, Article I, Section 8, provides that Congress shall "pay the debts and provide for the common defense and general welfare of the United States." These words are commonly interpreted as giving Congress broad ability to spend for the general welfare. By conditioning this spending power as a "carrot," Congress can use it to advance specific regulatory areas. Conditional grants to a nonfederal governmental unit for the performance of certain environmentally benign activities, for instance, are quite common.

Another significant national power is the Property Clause of the Constitution, which grants Congress the "power to dispose of and make all needful rules and regulations respecting the territory of other property belonging to the United States." This clause is of great importance because the federal government owns an estimated 732 million acres (or one-third) of the land area of the United States. This means that the federal government may, for example, compete with private businesses—which it could do (and has done) by selling electricity generated from dams built on federal property. It may also release stockpiles of federal resources in an attempt to influence market prices, as it has done with petroleum or uranium fuel. It is the Property Clause that gives Congress the power to protect the vast amount of natural resources contained in these public lands.

In addition to the Commerce Clause and taxing power, Clause 18 of Article I, Section 8, empowers Congress "To make all laws which shall be necessary and proper for carrying into execution the foregoing powers, and all other powers vested by this Constitution in the government of the United States, or in any department or officer thereof." This provision is known as the Necessary and Proper Clause.

The Necessary and Proper Clause, with the Commerce Clause, has been held to provide justification for broad congressional control of commerce that affects the environment. Essentially, the Necessary and Proper Clause provides Congress with the power to deal with matters beyond the list of specified federal concerns, as long as control of those matters will help Congress to be more effective in executing control over specified concerns. To illustrate, the U.S. Supreme Court ruled in 1978 that a federal statute that limited tort liabilities arising out of nuclear accidents to $560 million was necessary and proper to achieve the objective of encouraging the development of privately operated nuclear power plants.[4]

The treaty power is shared between two branches of the national government. The executive branch may make a treaty, but it must be ratified by two-thirds of the Senate (Article II, §2). A validly ratified treaty is almost equivalent to federal legislation. Consequently, when a conflict emerges between a valid treaty and a valid federal statute, whichever was enacted later controls under the principle that "the last expression of the sovereign must control." Moreover, a valid treaty is binding on the states, under the Supremacy Clause. The case that follows illustrates the effect of the Supremacy Clause on state statutes.

In the case of *Missouri v. Holland* (1920), Congress attempted to regulate the killing of birds within the United States.[5] This legislation was declared unconstitutional because it was not within one of the expressed congressional powers. Later, a valid treaty was made between the United States and Great Britain, governing migration of birds between the United States and Canada. The treaty prohibited the killing or capture of certain kinds of birds within the United States. The state of Missouri alleged that the act invaded rights guaranteed to it under the Tenth Amendment. The Supreme Court, in interpreting the Supremacy and Necessary and Proper Clauses, found that the treaty and its regulations were constitutionally valid, and did not violate any state's Tenth Amendment rights. Thus, even though a subject area might not otherwise be within one of the delegated powers of the federal government, if it falls within the scope of an otherwise valid treaty, it will be valid as a "necessary and proper means" of exercising the treaty power. Moreover, as in this case, the migration of wild birds is a federal concern, to be best dealt with by federal legislation. Thus, no Tenth Amendment rights of states will be allowed to stand in the way of a national problem.

The history and development of environmental policy in the United States may be summarized as follows. First, during the nineteenth century there was an increased interest in protecting the nation's rapidly vanishing wilderness. Second, there has been a great surge in environmental interest during the latter part of the twentieth century. The interest was sparked by an increase in pollution and resource depletion that came about as more Americans moved to the suburbs—and the economic prosperity and technological improvement that made those moves possible. Concomitantly, the federal government expanded its power over the regulation of environmental quality. Since the 1960s, the federal government has either supplemented state regulations or has taken primary responsibility for control of the quality of the nation's environment. In the words of one observer:

One of the most compelling themes to emerge from the decade of the 1960s was that the federal government must take a more pervasive role in solving what was beginning to be called "the environmental crisis." The limited partnership between the federal government and the states was insufficient to solve what was already being spoken of in global terms.[6]

DEVELOPMENT OF GLOBAL CONCERNS FOR PROTECTING THE BIOSPHERE

In the twentieth century, particularly after the 1960s, international awareness emerged that pollution and exploitation of natural resources were linked to broader economic issues, and urgent environmental problems required immediate attention. Events that helped to develop international environmental consciousness include the following:

- 1967—A series of oil spills culminated in the *Torrey Canyon* disaster, in which approximately seventy thousand tons of crude oil were spilled onto British and French seashores.

- 1984—A poisonous cloud leaked from the Union Carbide pesticide plant and killed more than two thousand people in Bhopal, India.

- 1985—The discovery of an "ozone hole" in the stratosphere over Antarctica.

- 1986—Nuclear power plant disaster at Chernobyl in the Ukraine spewed more radioactive material into the atmosphere than was released by the atomic bombs that destroyed Hiroshima and Nagasaki in World War II.

Add to this the deforestation of much of the remaining forests, the extinction of thousands of species worldwide, and the possibility of global climate change during the twenty-first century.

Exploiting nature is not of modern vintage. Three hundred years before the birth of Christ, once-fertile lands along the coast of Turkey became deserts principally through overusage. Historians and scientists are beginning to recognize the significant role natural resources played on world civilizations. Recent evidence suggests that the rise and fall of civilizations can be, to a large degree, attributed to environmental factors. For example, lead poisoning from lead pipes and drinking vessels may have killed numerous members of the ruling class of the Roman Empire, thus contributing to the fall of ancient Rome.

The decline of many civilizations can largely be explained by two factors: first, the intensity of pressure by the civilization on the resource base; second, the attitudes of the inhabitants toward the rational use of the resource base, so as to make it sustainable. The phenomenon of ecological deterioration has been summed up by one ecologist:

The span of history from 5,000 B.C. to 200 A.D., which we know primarily as the period of great civilizations—Sumeria, Babylonia, Assyria, Phoenicia, Egypt, Greece, and Rome—was also a period of unprecedented environmental disturbance. We tend to concentrate our attention on the superb achievements of these civilizations in literature, art, government, and science, while we virtually forget their incompetence in land management. These golden civilizations prospered at the expense of their environments. They left a landscape which has never recovered, and a legacy to future civilizations which ushered in a period of dark ages lasting for more than a thousand years.[7]

We have only recently realized that human activity may be again inadvertently and irreversibly depleting its resource base. However, unlike the great civilizations of the past, some modern pollutants are pervasive and are bound to affect the integrity of the entire planet. Perhaps modern civilization will not follow the scenario of ancient civilizations, and will use international laws as a catalyst to protect the order of nature. Indeed, it would be ironic if "history repeated itself."

From an ecological perspective, there are two fundamental kinds of international problems. The first is due to overpopulation and occurs principally in developing nations. The second comes from increased resource consumption with its concomitant transboundary pollutants and is mostly associated with developed nations. To illustrate, the developed nations make up only 20 percent of the world's population yet are responsible for 80 percent of the Earth's consumption of resources and generation of pollution. The dilemma is that as developing nations pursue the goal of becoming industrialized, they too will contribute to the depletion of resources and to pollution.

Modern international concerns derive from the desire to preserve and protect the order of the ecosphere and to bring human activities into harmony with ecological ones. The following quote comes from government decision makers attending the 1989 Paris Economic Summit:

There is growing awareness throughout the world of the necessity to preserve better the global ecological balance. This includes serious threats to the atmosphere, which could lead to future climate changes. We note with great concern the growing pollution of air, lakes, rivers, oceans and seas; acid rain, dangerous substances; and the rapid desertification and deforestation. Such environmental degradation endangers species and undermines the well-being of individuals and societies.[8]

(Primary Document 28 provides excerpts from this summit.)

In going from the rhetoric of environmental concerns to the realities of environmental laws, the global community faces seemingly insurmountable problems. A few of them are worth reviewing. First, the human society plus its environment has an indivisible complexity. Societies are divided into villages, towns, cities, counties, and nations for political or jurisdictional purposes, but from an ecological perspective, nations are not independent systems. They are merely parts of a global mosaic of interdependent systems.

Different political systems greatly complicate the environmental decision-making process. Because pollution does not respect political boundaries, nations have been unable to control pollution originating outside their jurisdictions. Furthermore, people in areas where pollution originated often had little or no incentive to control the outflow because pollutants traveled downwind or downstream.

A second problem faced by the global community in developing international environmental laws is that scientific uncertainty hinders environmental decision making. Decision makers, such as legislators, public agencies with environment-related duties, and courts, need reliable environmental information in order to ascertain the probable consequences of their decisions. A major concern or limit to environmental jurisprudence is that scientists are not absolutely sure of how the varieties of pollutants affect human health and the ecosphere. The lack of a unanimous scientific opinion helps underscore the great complexity and limits the ability of environmental decision makers to regulate the environment or determine the scientific facts at issue.

To further illustrate the complexity of this problem, consider one of the world's most ubiquitous concerns, the uncertainties regarding global climate change. The possibility of global climate change induced by an increase of pollutants in the atmosphere is potentially the most important international issue facing humanity. Yet, decision makers have had a very difficult time dealing with this concern because of scientific uncertainty and because legal action on the warming of the Earth might well carry them too far from the fundamental problems of most of their constituents. (Chapter 3 considers the policy dilemmas that the United Nations faces in attempting to resolve the issue of global warming.) Because there is so much uncertainty as to the effects of humanity on the ecosphere, prudence—argue many decision makers—dictates erring on the side of caution.

A third major problem facing public decision makers is that legal changes tend to follow and depend on social changes rather than being anticipatory or progressive. For example, the Fourteenth Amendment was passed only after, and because of, the cataclysm of the American Civil War. The United Nations was created as a result of the devastation caused by World War II. Legal systems are reactive, and international systems are even more so. As described in Chapter 3, due to the long atmospheric life spans of the greenhouse gases and the inertia of the global environment, waiting for proof beyond a reasonable doubt about the occurrence of the "greenhouse effect" would be too late to prevent the consequences.

A fourth problem emerges from a fundamental dysfunction of international law. A unique feature of modern international law is that a treaty must result from the assent of all parties concerned. In addition, it is not possible to enforce international law unless all sovereign nations agree. This unanimous aspect greatly weakens proposed environmental policies because the draft must satisfy the interests of the most reluctant nation. Thus, maintaining the order of the ecosphere requires radical changes in the United Nations.

Today, not even the U.N. General Assembly comes close to being a true legislative body. Its resolutions have neither legal nor substantive binding power. Furthermore, the International Court of Justice hears only those cases referred to it by consenting governments or international organizations, and no executive authority exists to enforce its decisions. In addition, there is an underlying problem of convincing the global community that a proposed international law is not a scheme to erode their sovereignty on behalf of big-power economic and political interests.

Obviously, the list of reasons why nations will not always cooperate is long. One may add to this a variety of procedural inefficiencies, administrative delays and archaic legal technologies, conflicting loyalties, and different administrative traditions and languages. In an article on how various factors affect the development of environmental treaties, Robert Hahn and Kenneth Richards point out that the United States, before signing any international agreement, should cautiously consider the politics, economics, and underlying science that should be its concern. The United States should also consider that as a world leader it is more likely to be obligated to adhere to international commitments than most nations.[9]

Other issues besides the environment may be more important to individual countries. They may have relaxed pollution standards in order to encourage industrial development. Countries such as Croatia, Israel, Kuwait, Nationalist China, and Somalia have reduced standards because they are fighting for their survival as nations. In effect, the people in these nations are saying, "Give us peace, food, shelter, and clothing—then we will be concerned about the environment."

In spite of these shortcomings, global conferences have been convened and have produced a number of treaties and other agreements for addressing environmental problems. One of the earliest recorded attempts at such environmental cooperation was the Swiss government's proposal to establish an international commission to protect migratory birds in Europe (1872). Others that later followed were the European Convention Concerning the Conservation of Birds Useful to Agriculture (1902), and the Convention Relative to the Preservation of Fauna and Flora in Their Natural State (1933). Today, there are more than 170 environmental treaties, over two-thirds of which have been adopted since the 1970s. (The Primary Documents chapter presents some major international treaties and other agreements in the field of the environment.)

Prior to the 1960s, most treaties were marginal because of their limited scope and application. They did little to prevent pollution. Indeed, international pollution was increasing rather than decreasing. After the 1960s, however, many countries saw the need for a comprehensive review of the global commons and for recommendations on how to improve or at least protect them.

In 1968, at the initiative of Sweden, the U.N. General Assembly called for high-level intergovernmental conferences to address a range of environmental concerns. In 1972 the Stockholm U.N. Conference on the Human Environment was the first such meeting, and its developers had to set both the process and the focus for later global environmental conferences. The Stockholm Conference, which required four years of planning by a preparatory committee as well as a range of regional and scientific gatherings, was successful. The Conference, which included 133 countries (many of which had addressed their own environmental concerns for the first time as part of their preparation for the conference), adopted a declaration, a resolution on institutional and financial arrangements, and an "action plan" with 109 recommendations that can be grouped into five areas: (1) human settlements, (2) resource management, (3) pollutants, (4) education, and (5) development. (Document 25 provides excerpts from this historic conference.) The recommendations of the conference were submitted to and approved by the U.N. General Assembly at its fall 1972 session.

A key result of the Stockholm Conference was the creation of a broader action plan proposed for addressing global environmental issues. Stockholm also gave birth to an approach to environmental issues through large United Nations–sponsored global conferences that explored special concerns (see Primary Documents, Table 2).

These conferences revealed that during the last quarter of the twentieth century, a major shift had taken place in attitudes, interests, and priorities towards the environment. In the 1970s, in spite of an oil embargo and a few recessions, most nations were enjoying relative economic prosperity. The mood at that time was optimistic, with attention mostly aimed at controlling air and water pollution, the protection of natural environments and their species, and environmental actions taken on a national level.

The concerns expressed at conferences during the last two decades of the twentieth century have expanded due to the serious worldwide economic problems. One example of this is the idea of "sustainable development," a notion embracing both economic growth and environmental quality that are sustainable over the long haul. Richard Sandbrook, from the International Institute for Environment and Development, further elaborates:

The global approach to environmental problems has changed significantly since the United Nations Conference on the Human Environment in 1972. Then the primary concerns were about the externalized costs of industrialization in the "North" and poverty in the "South." Now there is far more concern about how development can include good environmental stewardship—how can development be sustainable?[10]

By the end of the twentieth century, international entities will represent the major vanguard of the environmental movement. These organizations have been designed to transcend parochial interests, and to address fundamental economic and other forces that influence the quality of the global environment. Special environmental units of international organizations have also been established. In addition, other regional entities, such as the Organization for Economic Cooperation and Development and the European Community, have established environmental units to make sure that environmental interests are a component of their organizational goals. (Many of these public and private organizations are shown in Table 3, Primary Documents.)

The shift in thinking also has included giving more attention to generating action at the local instead of the national level, and incorporating more participation of organizations outside government, such as business and industry. Thus, in this century there has been not only an evolution in the way humanity looks at nature, but also a great surge in global organizations to grapple with the depletion of resources and the pollution of the global commons.

The establishment and remarkable success of a wide range of nongovernmental organizations such as the World Wildlife Fund, the Defenders of Wildlife, and the Friends of the Earth cannot be overemphasized. Though the protection of environmental quality is the responsibility of government, private citizens, through nonprofit environmental organizations, have become increasingly effective in aiding developing nations over the past decade. This trend appears to reflect an awareness on the part of those involved that nongovernmental organizations, because of their direct, person-to-person approach, are especially suited to the task of addressing environmental concerns at the grassroots level. To illustrate, technology relevant to small farms, such as small agro-forestry programs, which are actively sponsored by some nongovernmental organizations, is gaining in popularity as the most effective means of dealing with soil conservation on a sustainable basis.

Many environmental organizations operate international programs in a number of countries to encourage conservation of wildlife. Their goals are directed towards the preservation of a global network of natural biomes that represent the full biodiversity of life on Earth. (The activities of some of these organizations are summarized in platforms of nongovernmental environmental organizations.)

The development of a worldview came from the recognition that Planet Earth consists of integrated resources of land, air, and water; that environmental conservation and economic activities are substantially intertwined; and that without global collaboration a healthy, sustainable biosphere is not possible.

Looking back on the historical development of the environmental movement during the twentieth century, it is apparent that the world has partially shifted itself in a new direction. That is not to say that a Chernobyl-like accident could never reoccur, that people will never be exposed to a cancer-causing substance, that no species of wildlife will ever become extinct, that any natural environment will never be exploited, or that global climate change will not occur. There is a great deal of work yet to be done on these and many other environmental concerns. As the needs and goals of the global community change in the twenty-first century, new environmental issues no doubt will emerge to be reckoned with, and new environmentalists and organizations will arise to meet those challenges.

NOTES

1. Norman J. Vig and Michael E. Kraft, *Environmental Policy in the 1990s* (Washington, DC: Congressional Quarterly, 1994), 73.

2. Samuel P. Hays, *Beauty, Health, and Permanence: Environmental Politics in the United States, 1955–1985* (New York: Cambridge University Press, 1987), 5.

3. *City of Philadelphia v. New Jersey*, 437 US 617, p. 628 (1978).

4. *Duke Power v. Carolina Environmental Study Group, Inc.*, 438 US 59 (1978).

5. *Missouri v. Holland*, 252 US 416 (1920).

6. Jacqueline V. Switzer, *Environmental Politics: Domestic and Global Dimensions* (New York: St. Martin's Press, 1994), 14.

7. Charles H. Southwick, "Environmental Impacts of Early Societies and the Rise of Agriculture," in C. H. Southwick, ed., *Global Ecology* (New York: Oxford University Press, 1996), 209.

8. 1989 G7 Paris Economic Summit, *Weekly Compilation of Presidential Documents* (July 16, 1989), 1105.

9. Robert W. Hahn and Kenneth Richards, "The Internationalization of Environmental Regulation," *Harvard Journal of International Law* 30 (1989): 421–46.

10. Richard Sandbrook, "Towards a Global Environmental Strategy," in C. C. Clark, ed., *Environmental Policies* (Dover, NH: Croom Helm, 1986), 302.

2

The Concern for Our Vanishing Wilderness

The history of the environmental crisis can be studied by looking at the chronology of events that led to society's awareness of key environmental problems. Many environmental scientists consider this crisis to be a combination of interconnected problems, such as the vanishing wilderness, pollution, and overpopulation, all of which may have additive and synergistic effects. However, underlying these problems is a more subtle issue. It concerns the development of a human society that is profoundly different from the one in which we live: the "self-sustainable society."

This chapter looks at the first major concern of the environmental movement—the "value of nature phase." During this stage, the loss of wilderness (or ecologically speaking, the biosphere) becomes so great that concerned citizens have begun conservation movements to sequester and preserve extraordinary natural regions. An examination of species extinction suggests the basic contours of the problem of maintaining biodiversity. Because the loss of wilderness is partly due to the Judeo-Christian attitude that God created nature solely to serve humankind, a worthwhile review of some of the principal ethical traditions of Western philosophy reveals how those traditions have led to our vanishing wilderness.

CONDITIONS AND TRENDS IN BIODIVERSITY

The term *biodiversity* refers to both the variety and variability in species and the genes that they contain. The term *biosphere* is a more inclusive term used to define the parts of the Earth that these species inhabit. Both biodiversity and the biosphere are vanishing on Earth. The rate of species extinction has a significant impact on society, as well as on the entire biosphere. For example, genetic diversity better enables species to adapt to changing environmental conditions. Species diversity is the key to such fundamental ecological concepts as the food web and ecosystem stability. Generally, the most-complex ecological communities, which are composed of many different species, are the most stable and therefore the most resistant to bioecological change. Simpler ecological communities are more fragile, being less able to withstand changes and survive.

Most scientists estimate that as of 1997, there are approximately fifteen to twenty million species on Earth. This rich variety of living things that exists on this planet is ultimately the source of society's food, clothing, shelter, and more. Natural ecosystems continually recycle air, water, and land. Society disrupts biodiversity by minimizing its options and quality of life.

In the biosphere, species are challenged by fluctuations in the physical environment, predation, parasitism, and competition for resources. Extinction results when species, highly adapted to one set of conditions, are unable to survive under new conditions. The history of the dinosaur attests to the eventual fate of many organisms—extinction. Since the origin of life, three to four billion years ago, researchers have estimated that 99 percent of all species that once existed have disappeared. This is because living organisms were unable to adapt to changes in the biosphere.

Although extinction is a natural biological process, it has been increasing at an alarming rate since the 1600s (see Figures 2.1 and 2.2). Approximately 150 mammals and birds have become extinct, following the patterns of the dodo bird, passenger pigeon, dusky seaside sparrow, eastern elk, plain wolf, and the Leon springs pupfish. Several scientists have predicted that by the end of the twentieth century, approximately one million of the world's species will vanish, and that another 30–50 percent will become extinct within five hundred years. Some pessimistic estimates suggest that extinction rates for all taxonomic species will be one to three species per hour.

The overwhelming rate of species extinction is mainly the result of the expansion of the human population, along with trophy hunting, economic harvesting, deforestation, wetland drainage, urbanization, agricultural clearing, pollutants, and the introduction of intrusive species into various ecosystems. Humanity is changing the environment and destroying the natural habitats too rapidly for most flora and fauna to adapt.

In the past, due to the open country and the vast amount of biosphere, the thought of the biosphere as a finite resource was inconceivable. However, as

Figure 2.1
Threatened and Endangered Wildlife (Number of Species),
1980–1993

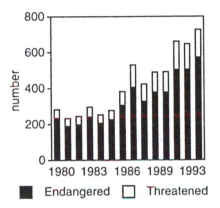

Source: Council on Environmental Quality. *Twenty-Fourth Annual Report*. Washington, DC: U.S. Government Printing Office, 1993.

Figure 2.2
Endangered Plants and Animals, 1993

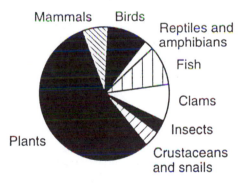

Source: Council on Environmental Quality. *Twenty-Fourth Annual Report*. Washington, DC: U.S. Government Printing Office, 1993.

the world now enters into the twenty-first century, recognitions of serious problems have called into question the carrying capacity of the biosphere that must support the world's human population. One of these problems is desertification, the process of converting arid and semiarid lands into veritable deserts. At the present pace of conversion, desertification could enormously decrease high agricultural productivity in both irrigated and nonirrigated lands. Desertification could also seriously decrease the carrying capacity of the arid lands that support human settlements and wildlife.

A second concern is the global trend of converting natural forest biomes to agricultural areas, urban centers, water resources, transportation, and other developments. If this trend continues, the number and size of natural forest biomes (tropical, deciduous, and coniferous forests) may be seriously reduced, thus eliminating the habitat that supports most of the growth of green plants, which in turn provide food either directly or indirectly to all living things.

Desertification

Desertification is the process of productive environments becoming increasingly desertlike. It is characterized by the lowering of water tables, a shortage of surface water, the salinization of existing water supplies, and wind and water erosion. In arid and semiarid regions the main cause of desertification is the overuse of land, resulting from overgrazing, overdrafting (mining) of groundwater for agriculture, poorly managed irrigation, the use of soils with poor drainage, and the choosing of crops and employment of agricultural practices that neglect soil conservation. Other variables leading to desertification are industrial development and rapid urban growth.

In her study on desertification, Sandra Postel, from the Worldwatch Institute, used the term "land degradation" interchangeably with desertification. She mentions the desertification scenario on rangelands:

Degradation on rangelands mainly takes the form of a deterioration in the quality and, eventually, the quantity of vegetation as a result of overgrazing. As the size of livestock herds surpasses the carrying capacity of perennial grasses on the range, less palatable annual grasses and shrubs move in. If overgrazing and trampling continue, plant cover of all types begins to diminish, leaving the land exposed to the ravages of wind and water. In the severest stages, the soil forms a crust as animal hooves trample nearly bare ground, and erosion accelerates. The formation of large gullies or sand dunes signals that desertification can claim another victory.[1]

Presently, desertification of the Earth's land surface is increasing at an estimated rate of 50,000 square kilometers (19,305 square miles) per year. If this continues, 35 percent of the world's usable land surface will be lost by the end of the century. However, the desertification problem is not a modern phenomenon. Three hundred years before the birth of Christ, once-fertile lands along the coast of Turkey became deserts, principally because of overuse by their human inhabitants.

Arable land per capita is declining while the world population is expanding rapidly. This rapid loss of productive lands can have severe repercussions for both humankind and the environment. For instance, the arid western parts of the United States are rich and productive; however, they are also undergoing desertification, which will seriously threaten their long-term productivity. In developing nations, desertification causes great human misery. Desertification in parts of Africa and western Rajasthan (India) has led to malnutrition or undernourishment for millions of people. In the past fifty years in Africa alone, 650,000 square kilometers of land bordering the Sahara, once suitable for agriculture and grazing, have become barren deserts. In the southern rim of the Sahara, the Sahelian drought of 1968–1974 caused thousands of people and tens of thousands of animals to die. The Sahara expanded, creeping southward by five degrees latitude. In the last few years, another severe drought in the Sahel has caused the death of thousands of people, decimation of livestock and wildlife, and further desertification.

There are remedies that can aid developing countries with this problem. Some scientific-technical remedies to decrease desertification problems include improved management practices (e.g., replanting or reduced grazing) and certain techniques that keep the water closer to the upper portion of drainage valleys (watersheds) in order to restore previous plants and hydrology. However, if degradation reaches the stage of erosion, where soil and nutrients are irreversibly lost, these lands may never be reclaimed at a reasonable cost.

Moreover, in developing nations, the scientific-technical remedies are not enough. In these regions, aggravating the situations are fundamental socioeconomic problems of unequitable land tenure, lack of employment, and rapidly growing human populations. The rural poor must gather firewood because they cannot afford coal, gas, or kerosene. Their only option is to overharvest arid and semiarid areas for fuel and food. This is caused by the scarcity of suitable land and capital, which are essential to providing for their families. The concern for protecting these vulnerable areas, although they will not be affected for a decade or two, appears understandably small to these peasants compared with the value of feeding their families today; therefore, conservation is left to be a concern only for those few altruistic individuals who are not starving. As Don Paarlberg, the first coordinator of Food-For-Peace Program, quoting from an old Byzantine proverb, states, "He who has bread has many problems; he who lacks bread has only one problem."[2] Consequently, many environmentalists argue that the principal impetus for protecting arid and semiarid areas in poor regions must come from international aid agencies and from developed nations through scientific and economic aid programs.

Deforestation

The most important terrestrial natural ecosystems are tropical rain forests, deciduous forests, coniferous forests, grasslands, tundra, chapparals or scrubs, and deserts. Likewise, these natural areas all suffer identical problems of conversion to other uses and/or damage due to pollution. Forests are especially under intense pressure. Andrew Goudie, a professor of geography at Oxford University, believes that the "deliberate removal of forest" is "one of the most longlasting and significant ways in which humans have modified the environment, whether achieved by fire or by cutting."[3]

The U.N. Food and Agriculture Organization has reported that between 1980 and 1990 tropical forests had been destroyed on an average of 15.4 million hectares (0.8 percent) per year.[4] At the present rate of deforestation, the tropical forests will be gone by the end of the twenty-first century. Tropical forests cover just 6 percent of the Earth's surface, but more than 50 percent of all species dwell in them.

Tropical forests are lost each year through the combined activity of deforestation, fuelwood gathering, and cattle grazing. One of the most damaging agricultural activities is slash-and-burn agriculture. This procedure involves cutting down a small patch of forest, burning the native vegetation, using the ash as fertilizer, and raising crops for a year or two in the ash-enriched soil. When the nutrients in the ash are depleted, the farmer and his family move on, leaving the recently used cropland fallow until wild plants have grown again. Revegetation by wild plants restores soil nutrients but this occurs decades or centuries later (depending on the nature of the forest being cut). Shifting agriculture can be practiced indefinitely where soils and terrain are favorable and population density is low. However, shifting cultivators in many tropical forests include shortening fallow periods and using the land intensively. The fundamental justification for increased agricultural use is the loss of forest availability to traditional cultivators, because of competing demands such as large plantations and urbanization, and the migration of peasants due to the lack of land or jobs elsewhere.

Deforestation also occurs when people routinely destroy trees and clear whole forests in order to obtain firewood, clear land for cattle raising, or cut timber for exportation. Due to the relatively high cost of other forms of energy (for example, coal, kerosene, gas), firewood is usually the most important fuel in developing nations. In some of these nations, fuel wood provides more than 90 percent of the energy consumed. Timber harvesting (especially in tropical Asia) and deforestation for cattle raising (especially in Latin America) have steadily gained value since World War II.

Presently, deforestation can be reversed in many regions through tree planting programs, removing land from agricultural use, and employing alternative nonfirewood sources of energy (such as geothermal, gravitational, wind, and solar). Similar to desertification in poor areas, the combination of economic poverty, land scarcity, and increasing human population has prevented reforestation programs from keeping up with forest losses. Therefore, argue environmentalists, remedies in poor nations need an integrated policy of scientific expertise and socioeconomic changes.

A 1977 United Nations Environment Programme report provided some ideas as to how a number of practical changes in behavior and techniques utilized in burning fuelwood might significantly conserve energy.[5] Normally, only 6 percent of the heat value is captured in a common wood cooking stove in Indonesia. However, when the stove is designed with a partial sunk-in pot receptacle, it conserves 20 percent of the energy; a newly designed pot saves another 30 percent; and drying the wood for several weeks before burning saves another 10 percent. The report suggests that if the overall efficiency improvements averaged 50 percent, a carefully managed ten-hectare (twenty-five-acre) firewood plantation cultivated with fast-growing plants could supply the fuel needs of an Indian village of one thousand families indefinitely. The problem again is largely socioeconomic, because most poor families do not have the money to buy newly designed stoves and pots. As the report suggests, the few who can afford these investments need to be convinced that in the long run they will result in a lower cost from the firewood saved.

In 1985 the World Bank published a report evaluating the population-sustaining capacity of seven West African countries.[6] The study centered on determining the carrying capacity of these countries limited by fuelwood and food supplies. The report concluded that in regions where rainfall is lowest, sustainable agricultural and fuelwood yields were well below those that the regions demanded. Surprisingly, fuelwood emerged as the critical factor.

The report indicated that the actual population of all seven nations in 1980 was thirty-one million. The population obviously already had exceeded the ideal twenty-one million that could be sustainably supplied with wood resource yields. Unfortunately, with the population predicted to grow to fifty-five million by the year 2000, the total degradation of the land is bound to increase with its concomitant effect on wildlife and on the ability to support human needs.

Deforestation is also occurring in temperate and coniferous forests of developed nations. Since the creation of the United States, the history of land utilization has fundamentally affected the pattern of forest distribution in the present-day United States. In fact, only scattered stands of trees remain of the original majestic forests that covered the eastern parts of northern America, the West Coast, and the western interior parts of the United States. This is due to deforestation for farms, harvesting timber, urbanization, and conversion for other uses. Today, in the Pacific Northwest, logging of old-growth forests is a major concern. In some areas, the forests are substantially degraded due to air pollution, inefficient harvesting, and lack of reforestation. In some parts of eastern Europe, deforestation is as high as 50 percent. Because more than one-third of the forest is dying as a result of acid rain, Germany is concerned with its possible effects on the Black Forest.

ORIGINS OF THE CONCERN FOR PROTECTING BIODIVERSITY

There are many justifications for protecting the biosphere and its biodiversity. They can be grouped into two broad categories: anthropocentric and ecocentric.

Anthropocentric Ethics

Predicated on anthropocentric ethics, the conservation of biodiversity is justified from the standpoint of its benefit to humankind. The root of these ethics can be partially traced to the Judeo-Christian tradition that God created nature solely for human benefit.

In ancient Greece and Rome, the relationship existing between the gods and humankind differed dramatically from those existing in the Jewish and Christian religions. In the Greco-Roman world, it was believed that the whole environment was inhabited by different deities and spirits. Every stream and mountain had a god or goddess. Indeed, some of these gods, such as Pan, Artemis, and the Roman Diana, were predominantly gods of nature—protectors of wildlife.

Such beliefs seemingly promised a very fortunate situation for the environment, but, in fact, recent archaeological and geological evidence indicates that the Greeks and Romans did not live in harmony with their environment. These ancient societies may have crossed vital thresholds of natural and social stability, which led to the "decline of classical civiliza-tion." In the words of the historian Edward Anson:

Elements of ancient Greco-Roman civilization which should have protected the en-vironment of the Mediterranean world from harm were negated by other elements. For example, while there did derive from primitive Greek and Roman religion a general reluctance to tamper with nature for fear of offending the gods, Greco-Roman practically united with universal human greed to render this apparent safe-guard inoperative. Certain places, *temene*, came to be regarded as reserved for dei-ties, with the rest of the countryside, in consequence, being considered less sacred and more subject to human whim.[7]

Environmental deterioration was further exacerbated with the emer-gence of Judeo-Christian ethics. Under the Jewish and Christian religions, humans divorced themselves from nature and entered into an exclusive re-lationship with God. This worldview generally showed a lack of regard for nature. The environment then became a source of property for humanity and was exploited to meet its needs.

This anthropocentric situation continued into the seventeenth century, when Francis Bacon put forth the philosophy that science and technology could be used to control and even dominate nature for the benefit of hu-mans. Even among scientists, this notion was unquestionably accepted. Swedish biologist Carolus Linnaeus, founder of the binomial classification system used by modern biologists, reasoned that God created "nature's economy" solely to serve the human economy. According to Linnaeus,

All the creatures of nature, so artfully contrived, so wonderfully propagated, so providentially supported . . . seem intended by the Creator for the sake of man.[8]

It is possible to see the industrial revolution as a logical consequence of the Baconian philosophy. Until recently, the pleasure-loving Western soci-ety had mostly ignored or overlooked its responsibility to the environment. Today, however, a growing awareness of humankind's mutualistic relation-ship with the biosphere has revealed how the survival of society as we know it depends on the continued existence of the natural world.

In 1967 Lynn White, a historian of science and technology, wrote a fa-mous essay entitled "The Historical Roots of Our Ecologic Crisis," in which he argued that the responsibility for the environmental crisis is rooted in the traditional Judeo-Christian proposition that nature exists to serve hu-mankind. As summarized by White:

More science and more technology are not going to get us out of the present ecologic crisis until we find a new religion, or rethink our old one.... Both our present science and our present technology are so tinctured with orthodox Christian arrogance toward nature that no solution for our ecologic crisis can be expected from them alone. Since the roots of our trouble are so largely religious, the remedy must also be essentially religious, whether we call it that or not.[9]

The anthropocentric attitude is predicated on this relevant passage in the Bible:

And God said, Let us make man in our image, after our likeness: and let them have dominion over the fish of the sea, and over the fowl of the air, and over the cattle, and over all the earth, and everything that creepeth upon the earth. So God created man in his *own* image, in the image of God he created him; male and female he created them.[10]

Under the umbrella of the anthropocentric ethics, economic argument rests on the belief that genetic diversity possesses the potential for vast economic benefits for humankind. Conservationists using the economic argument point to such facts as: Food products from wildlife include fisheries, which in 1989 yielded one hundred million tons of food globally; and domestic organisms account for 12–32 percent of gross domestic product worldwide. The advocates also warn that by reducing biological diversity, humanity threatens to squander its greatest renewable natural resource for nutrients, energy, fibers, wood, wood preservatives, drugs, beverages, gums and related substances, essential oils, resins, tannins, cork, dyes, fatty oils and related substances, latex products, aesthetics, and countless other economic benefits. Biodiversity, advocates add, also attracts recreation and tourism to unique regions throughout the world.

Another subcategory of the pragmatic justification for biodiversity conservation is that many important medical drugs come from, or were first found in, wild species. Living things are a significant source of medicinal chemicals for traditional medicine, which is practiced by 80 percent of the people in developing nations. In addition, nature energizes the development of modern pharmaceuticals. For example, more than three thousand antibiotics (including penicillin and tetracycline), which account for one-quarter of all prescriptions filled in the United States, were first derived from microorganisms.

Conservationists using the anthropocentric justification also point out that humans still need nature to maintain their life-support system. Microorganisms carry out chemical reactions that affect the chemical composition of the atmosphere. For example, microscopic algae are so numerous that they contribute about 80 percent of the total photosynthesis that occurs on Earth. Consequently, they are responsible for most of the oxygen in the atmosphere. Bacteria, to further illustrate, convert nitrogen in the air to chemical compounds that can be used by plants.

In their natural state, forests regulate water flow, prevent soil erosion, and may have a major influence on regional and world climate. Some scientists have serious concerns that extensive loss of forest cover may magnify the increasing concentrations of carbon dioxide in the global atmosphere. Thus, deforestation will contribute to unprecedented human-caused changes in the world climate (discussed in more detail in Chapter 3).

In addition, natural forests provide recreation and unique scenic beauty while at the same time serving as the basis for natural communities that provide life support to organisms (including people). As mentioned, one vital by-product of plant photosynthetic activity is oxygen, which is essential to human existence. In addition, forests remove pollutants and odors from the atmosphere. The wilderness is highly effective in metabolizing many toxic substances. The atmospheric concentration of pollutants over the forest, such as particulates and sulfur dioxide, are measurably below that of adjacent areas (see Figure 2.3).

In view of their ecologic role in ecosystems, the impact of species extinction may be devastating. The rich diversity of species and the ecosystems that support them are intimately connected to the long-term survival of humankind. As the historic conservationist Aldo Leopold stated in 1949,

Figure 2.3
Forest Effectiveness in Metabolizing Toxic Substances

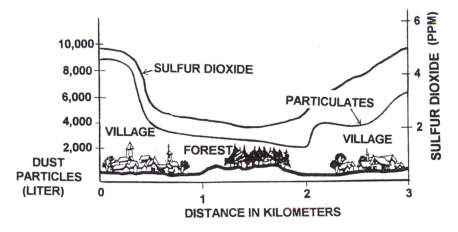

The outstanding scientific discovery of the twentieth century is not television or radio but the complexity of the land organisms. . . . To keep every cog and wheel is the first precaution of intelligent tinkering.[11]

An endangered species may have a significant role in its community. Such an organism may control the structure and functioning of the community through its activities. The sea otter, for example, in relation to its size, is perhaps the most voracious of all marine mammals. The otter feeds on sea mollusks, sea urchins, crabs, and fish. It needs to eat more than 20 percent of its weight every day to provide the necessary energy to maintain its body temperature in a cold marine habitat. The extinction of such keystone or controller species from the ecosystem would cause great damage. Its extinction could have cascading effects on many species, even causing secondary extinction.

Traditionally, species have always evolved along with their changing environment. As disease organisms evolve, other organisms may evolve chemical defense mechanisms that confer disease resistance. As the weather becomes drier, for example, plants may develop smaller, thicker leaves, which lose water slowly. The environment, however, is now developing and changing rapidly, but evolution is slow, requiring hundreds of thousands of years. If species are allowed to become extinct, the total biological diversity on Earth will be greatly reduced; therefore, the potential for natural adaptation and change also will be reduced, thus endangering the diversity of future human life-support systems.

The endangered species also may possess other characteristics that may make them valuable to humans. For example, humans existed several million years before research groups found useful substances from invertebrate animals including potential heart drugs from fireflies, a cockroach repellent from millipedes, and shark repellent from a marine mollusk. This genetic diversity is also essential to the development of new crops and livestock. For example, disease-resistant barley (*Hordeum vulgare*) has been developed by crossbreeding primitive barley varieties with newer varieties and by subsequent artificial selection. No one can predict which species will provide valuable material benefits in the future. Humankind's well-being depends on the preservation of as many species as possible.

As previously indicated, the tradition of anthropocentric ethics derives to a large degree from mutualistic and stewardship notions that there is a special and close association between humans and the natural world—that is, that the Earth provides a habitat and resources for humanity. Meanwhile, humanity should be careful not to destroy nature (i.e., "don't bite the hand that feeds you" or "live off the interest not the capital"). It follows that certain prudence is required when considering activities that may upset the stability of nature.

Ecocentric Ethics

With the concern for protecting the vanishing biosphere, there are ethics of yet another sort—namely, provisions aimed at maintaining nature, because species have a value that transcends economic, medical, or survival. Environmental measures predicated on this notion have been called ecocentric, biocentric, or preservation ethics (depending on the author's description). Whichever term is used, under this notion, the concern for protecting the order of nature derives mainly from social values of preserving and protecting nature, not because of some anthropocentric reason, but merely because nature has rights independent of human existence. Ecocentric ethics are based on moral, ethical, or religious reasons, and therefore, are more altruistic than anthropocentric ethics.

In both anthropocentric and ecocentric ethics, society is making a substitute judgment as to the fate of nature. The former, however, is more "subjective" in that we are deciding the value of nature from a human perspective—"our best interest." The latter is more "objective" or rational in that the fate of nature is decided from an ecological perspective—"nature's best interest."

Ecocentric ethics derive in part from concepts of natural law. Beginning in the sixteenth century, Western scholars first differentiated natural law (*jus naturale*) from positive law or the law practiced by nations (*jus gentium*). Natural law was the rights possessed by all humankind. The ultimate source of law was in nature rather than in the rules of society. Positive law accepts the notion that nations actually have provided more-relevant norms for the conduct of international relations. Baron Montesquieu, in his classical book *L'Esprit de Lois* (1748), argued that natural laws predated society and were superior to those of religion and of the state.

The notion of ecocentric ethics has gained prominence recently among many environmentalists, but its origins go back centuries. St. Francis of Assisi (1181–1226) espoused a philosophy akin to the ecocentric ethic. He specifically considered all animals as integral components of divine creation. According to him, wildlife has a right to exist independent of any human purpose. During St. Francis's century, however, such concepts were largely ignored or chastised.

The transcendentalist Henry David Thoreau (1817–1862) also professed a variety of ecocentric ethics. Thoreau built a cabin on Walden Pond in Massachusetts and lived a simple life for two years. While there, he recorded his experiences and reflections in his masterwork *Walden*, which presented the natural world as home to all living beings. More a philosopher than an activist, he viewed "nature" as a single organic entity with all living organisms related to each other. In *Walden* and other writings, Thoreau urged his readers to recognize and learn to live within environmental guidelines. He summed up his disposition and portentous feelings:

But most men, it seems to me, do not care for Nature and would sell their share in all her beauty, as long as they may live, for a stated sum—many for a glass of rum. Thank God, men cannot as yet fly, and lay waste the sky as well as the earth! We are safe on that side for the present. It is for the very reason that some do not care for those things that we need to continue to protect all from the vandalism of a few.[12]

In the twentieth century, one of the most influential figures in the development of ecocentric ethics was the American naturalist John Muir. He regarded wildlife as sacred and argued that nature has rights to exist and follow its own path independent of humanity. Muir believed that all species have a fundamental right to habitat protection. In his words:

The world, we are told, was made for man—a presumption not supported by all the facts. . . . Nature's object in making animals and plants might possibly be first of all the happiness of each of them. . . . Why should man value himself as more than a small part of the one great unit of creation?[13]

Forester Aldo Leopold later adopted a philosophy akin to Muir's. Leopold's ecological values are immortalized in his most famous book *A Sand County Almanac* (1949), where he views the sum total of life on Earth, together with its physical environment, as a kind of "living organism." Leopold regarded the killing of any wildlife as unjust not only because of the injury and suffering inflicted on the animal, but because it harmed nature as a whole. Under this notion, nature itself has intrinsic moral values, and society should develop a love for the land and its wildlife; this love is manifested through our stewardship of the land. Moreover, he argued that modern society has a duty to care for all parts of the biosphere, including its species. Given the impact of modern society's misuse of technology, Leopold added that nature cannot take care of itself—it is unbalanced.

More recently, the "deep ecology" movement, which includes an assortment of loosely related environmental values, has emerged. The term *deep ecology*, coined by the Norwegian Arne Naess, refers to a holistic and nonanthropocentric worldview that rejects the human-versus-environment distinction. Naess distinguished it from what he called "shallow ecology," or the environmental movement that focuses primarily on anthropocentric values. Although similar to Leopold's and Muir's notions in many of their facets, deep ecologists tend to be even more radical in their ecocentric perspectives. Some deep ecologists are extreme egalitarians calling for radical reforms such as alternative lifestyles, intimate contact with nature, and direct personal action to protect wildlife. In some cases, in seeking inspiration and divine revelations from nature, deep ecologists are imbued with spiritual, almost religious fervor.

The ideas of Muir, Leopold, and the deep ecologists have been important in the development of "Gaianism." The concept of Gaia derives from the recent theory of James Lovelock and Lynn Margulis who argued that the entire Earth's ecosphere operates as if it is a single superorganism. According to this hypothesis, the Earth is alive in the sense that its components have evolved in ways that help maintain the delicate balance necessary for life. Most professional ecologists agree that the environment modifies biota and that biota modify the environment to some degree, especially on a local scale. However, the idea that Earth's biota monitor and manipulate the physical environment to meet their needs has few scientific proponents. To some extent, this is due to the difficulty of testing the Gaia hypothesis. How did Gaia evolve? The process of natural selection occurs without a plan or purpose as different organisms compete for resources.

According to the theory of evolution by natural selection, because members of the same species differ from each other in their ability to compete for resources, avoid predators, and various such skills, those that are most suited for their environment are the ones most likely to survive and produce offspring. That is, "by survival of the fittest," only individuals with adaptive traits live long enough to have offspring. These differences in adaptiveness among individuals of the same species are due to genetic differences that may be transmitted from generation to generation. Over many generations (evolutionary time) differential reproduction among members of the same species with different genetic constitutions would lead to the species having a different genetic makeup.

Under the concept of Gaia, because competition between different Earths is missing, if Gaia evolved, it must have been through another mechanism of evolution besides natural selection. Consequently, most scientists are reluctant to accept the Gaia hypothesis. Nevertheless, one might consider the notion of Gaia as a useful metaphor for the interconnectedness of nature. This alternative point of view reminds us that although society tends to think of the Earth as a stable planet, we as living organisms have changed it in many ways. In that sense, the Gaia concept is no different than the biosphere concept. (The term *biosphere* was used by the Russian ecologist Vladimir Vernadsky in 1929 to refer to the layer of the Earth containing living organisms.)

The dichotomy between ecocentric and anthropocentric philosophies and practices also exists in non-Western cultures. A brief look at the environmental values of Native American culture, Buddhism, Hinduism, and Confucianism illustrates the difficulties in justifying the value of nature.

Many Native Americans revered nature, considered it sacred, and sought harmony with their environment. Although Native Americans differed greatly in their cultures and ways of using nature, like many precapitalist people, they respected the natural world. Some sense of what became a worldview, common among many Amerindian groups, was expressed in the following excerpt from a speech attributed to a Cherokee:

In the language of my people, Ani Yunwiyah, or Cherokee as we are called, there is a word for land: Eloheh. This same word also means history, culture, and religion. We cannot separate our place on earth from our lives on the earth nor from our vision and our meaning as a people. We are taught from childhood that the animals and even the trees and plants that we share a place with are our brothers and sisters.

So, when we speak of land, we are not speaking of property, territory, or even a piece of ground upon which our houses sit and our crops are grown. We are speaking of something truly sacred.[14]

According to Buddhism, a pan-Asian religion and philosophy, the objective of religious practice is to be rid of the delusion of ego, thus freeing oneself from the fetters of this mundane world—an absence of craving for material wealth. Individuals who are successful in doing so are said to have overcome the round of rebirths and to have achieved "Enlightenment"—the highest of all goals. Thus, under Buddhism, society's living standard is measured by the way it maximizes human well-being, not by its gross national product.

Yet another sharp contrast to the Judeo-Christian tradition is found in Hinduism, the socioreligious practice of the South Asian people (who are known as Hindus). Common characteristics of Hinduism include vegetarianism as well as respect for and consideration of all living organisms. A keystone of the ethics behind Hinduism is the belief in the unity of life.

Confucianist values represent the way of life followed by the Chinese people for well over two thousand years. According to this value system, private interests must yield to the public good. Thus, under Confucianism, the interest of the whole biosphere is greater than the human interest. There is a prophetic aspect to Confucianism's espousal of simplicity, nonviolence, and the absence of craving of material goods. Confucius said:

The superior man is concerned with virtue; the inferior man is concerned with land. . . . The superior man understands what is right; the inferior man understands what is profitable.[15]

It is easy to see how such worldviews among Native Americans, Buddhism, Hinduism, and Confucianism could be much more environmentally benign than many Western cultures. Unfortunately, in going from the philosophy of environmental values to the realities of biodiversity conservation, these non-Western cultures also have their problems. Although many Native Americans sought a natural balance, evidence indicates that when humans invaded the Americas about thirty thousand years ago, they hunted and exterminated fifty-six species of large mammals including camels, elephants, horses, a giant ground sloth, the sabertooth tiger, and a lion. Moreover, the Mayan civilization collapsed shortly after A.D. 800, possibly due to the overcultivation of fragile lands.

Buddhism also contributed to environmental problems. It introduced the practice of cremating the dead to China. From A.D. 1000 to 1400, cremation was extremely common in parts of China. This practice, however, caused a timber shortage. In addition, the traditional views of their non–Judeo-Christian beliefs have not prevented twentieth-century China and India from having an overpopulation problem with a concomitant effect on the biosphere. Indeed, by the year 2000, these two nations combined are projected to have 2.2 billion people—40 percent of the world's population.

POLICIES AND PROGRAMS FOR PROTECTING THE BIOSPHERE

In general, one of the problems in deciding the value of nature is that environmental values are subjective and prescriptive, instead of objective and descriptive. The normative nature of environmental values places constraints upon environmental legislation because different lawmakers will have different subjective notions of environmental policy. If environmental decision makers cannot decide on an environmental goal and the means to address the problem, environmental laws will be difficult both to enact and to enforce.

Most environmental lawsuits involve the measurement of nature's value and the balancing of it against purely economic-legal-social-political considerations. The government, for example, may license a nuclear power plant that is opposed by neighboring residents who fear a nuclear accident. The Tennessee Valley Authority may propose to build a power-generating dam that is supported by developers, but farmers and environmentalists fight it because of its effect on irrigation and endangered species.

Environmental values cannot be treated as independent of the political, economic, and social conditions, or the religious ethics that sustain those values. In fact, it has been suggested by at least one observer that "environmental policy-making often turns out to be a battlefield for religious conflict."[16]

Consequently, environmental policies or values must be clearly self-evident or at least intuitively attractive. As the founder of modern ecological ethics Aldo Leopold wrote:

The "key-log" which must be moved to release the evolutionary process for an ethic is simply this: quit thinking about decent land-use as solely an economic problem. Examine each question in terms of what is ethically and aesthetically right, as well as what is economically expedient. A thing is right when it tends to preserve the integrity, stability, and beauty of the biotic community. It is wrong when it tends otherwise.[17]

If an agreement with this ethical paradigm cannot be reached, then environmental objectives based on reasoning cannot proceed any further.

Unfortunately, the value phase is purely subjective, not scientific. The natural sciences describe the structure and dynamics of the environment; they do not prescribe the environment. This constraint places a limit on the use of scientific knowledge in evaluating the desirability of environmental goals—that is, species conservation. Science cannot determine if the extinction of a particular species is good or bad, moral or immoral, ugly or beautiful. For example, to a male gorilla, a human female holds no attraction as a mate. "Beauty is in the eye of the beholder."

As already noted, scientific principles are descriptive, not prescriptive. They do not say how things should be; instead, they say how things are and probably will be. This objective test precludes the acceptance of subjective knowledge. A scientist may testify as to what was, is, and will be; however, a scientist ventures beyond science if he/she testifies as to what should be. What "should be" is a normative, value-endowed judgment. For example, an ecologist testifying about deforestation can cite the evidence demonstrating that removal of the forest may accelerate the extinction of a certain endangered species, but the ecologist judging such extinction as wrong has left the realm of science for morality, philosophy, or politics. Perhaps, due in part to the descriptive nature of the science of ecology, there is a constitutional provision of the Ecological Society of America prohibiting it from lobbying. According to its Constitution:

No substantial part of the activities of the Ecological Society of America shall consist of carrying on propaganda, or otherwise attempting to influence legislation. . . . The Ecological Society of America shall not participate in, or intervene (including the publishing or distributing of statements concerning) political campaigns on behalf of (or in opposition to) any candidates for public office.[18]

It is not surprising then that most professional ecologists are reluctant to engage in the overt advocacy of a particular environmental issue.

The fact that ecology cannot be a source for all environmental values does not mean that it cannot be considered to be a foundation for some values or, indeed, for environmental values per se. Mark Sagoff, the director of the Institute for Philosophy and Public Policy at the University of Maryland, suggests that "the law calls upon ecology for guidance not just to make environments more productive, but also to protect ecosystems for their intrinsic natural qualities." He went on to emphasize this by saying that "ecologists then serve policy makers not only by helping achieve given objectives, to increase the profitability of farms and fisheries, for example, but also by helping them decide what their objectives should be, i.e., what to preserve and why."[19]

Given that ecology is descriptive and environmental values are normative, the great challenge now facing society is how to incorporate ecological principles into a value-laden environmental decision-making process. Perhaps there are cases, such as when action or inaction might create an unstable human-environment system, where science can participate directly in making public policy.

To illustrate the direct influence of science on public policy, consider the history of laws regulating health and the environmental laws protecting wetlands. After Jonas E. Salk developed the vaccine for poliomyelitis, some governments mandated that all children entering, or already in, school must be vaccinated. In the same fashion, once ecologists perceived that estuaries are generally productive and that marine life concentrates there because of high nutrient conditions, they were able to form the foundation for most laws aimed at the protection of these coastal wetlands.

Moreover, according to U.S. Federal Rules of Evidence, once a particular scientific principle has become sufficiently well-established (i.e., normally accepted among scientists), the legal system no longer needs proof (expert testimony) of the underlying basis of the theory. In other words, theoretical justification of a scientific theory is not as important as whether or not the theory is widely accepted by scientists.

Consequently, it seems that once scientists are capable of determining for a fact that the maintenance of the integrity of the biosphere is beneficial to society, governments may begin to enact laws protecting the biosphere and its species. The rationale again will be the long-accepted vital social objective of the state's compelling interest of protecting the health, safety, and welfare of the public. In effect, environmental laws are intervening with the values of individuals directly to promote the greater good of society, and indirectly to promote the individual's greater good.

Ultimately, most decisions involving environmental concerns are based on society's dominant cultural values, although science may be critical in shaping public policy. These include the values of public decision makers, nongovernmental leaders, media, and socioeconomic actors, as well as those with vested interest in the concerned activities. The remaining portion of this chapter views the historical background and future prospects of developing policies and programs to protect the biosphere.

Development of United States Policies
Concerning the Biosphere

When colonists first came to "new America" and migrated across successive frontiers, the land and its bounty seemed incapable of being depleted. America in the eighteenth and nineteenth centuries prided itself on being a nation of frontiers. The new Americans believed that settlers should develop and apply science to conquer and organize the savage wilderness. They believed that through the application of rigorous science, Americans would be able to create and rule over the advancing frontiers. In pursuit of these goals, Thomas Jefferson commissioned Lewis and Clark to set out on the first overland expedition (1804–1806) to the Pacific Coast and back. The expedition brought back valuable scientific information about the West and its inhabitants, thereby making the territory knowledgeable and accessible.

At the end of the American Revolution, as the new nation acquired new lands along the growing frontiers, large regions of the West became public domain. In 1812 the General Land Office (GLO) was established to supervise the disposition of this area. The GLO sold or granted much of the territory to the states for such public purposes as education and the expansion of railroads. The Homestead Act of 1862 was also significant in opening the

West to settlement because it provided a homesteader with 160 acres of free land for cultivation, development, and eventual ownership.

During these land transfers, protection of lands and their biodiversity to meet food and shelter needs were never expressed as goals. That was principally delegated to the new owners and developers. By the end of 1862, the national government had transferred 320 million acres, about 25 percent, of the 1.31-billion-acre public domain. As the U.S. population continued to grow, the disposition of public land also accelerated, and by the 1920s, public land transfer totaled almost 325 million acres, which was about one-third of the public land that remained in 1862.

The new Americans were concerned with the problems of living from day to day. They did not realize the impact they had on the land and its resources. In a land with such natural bounty, a need for restraint was unfathomable. The conservation movement developed in part from increasing concerns about the fate of the remaining public lands and the wildlife they contained.

A few individuals did express concern, as is evident in the "Charter of Rights" in Pennsylvania, which instructed the colonists there in 1681 that one acre of trees should be preserved for every five acres deforested. Furthermore, the concern for conservation goes as far back as when George Washington and Thomas Jefferson tried to conserve the wildlife on their properties. However, these individual acts of conservation did not slow the advance of frontiers. Indeed, public policy in the United States was geared toward the rapid settlement and development of successive "Wests" rather than concerns about threats to nature's abundance.

The first important manifestation of interest in conservation did not occur until 1864, with the publication of *Man and Nature* by George Perkins Marsh. Marsh wrote:

The ravages committed by man subvert the relations and destroy the balance which nature had established. . . . When the forest is gone, the great reservoir of moisture stored up in its vegetable mould is evaporated. . . . It is desirable that some large and easily accessible region of American soil should remain, as far as possible, in its premature condition, at once a museum for the instruction of the student, a garden for the recreation of lover of nature, and an asylum where indigenous tree, and humble plant that loves shade, and fish and fowl and four-footed beast, may dwell and perpetuate their kind.[20]

Geographer, lawyer, politician, and diplomat, Marsh was a man with many endeavors. He had a keen mind and prescient ideas on the limits of nature's bounty. Marsh's thought foreshadowed many of the fundamental

U.S. conservation efforts, leading into the twentieth century, as well as the course of U.S. involvement in global conservation efforts. He was, for example, one of the first to suggest that ancient civilizations declined as a result of environmental damage. Beyond lamenting past environmental damage, he described interactions among flora and fauna, and humanity's dependence on a balance of nature.

Among those affected by Marsh's warning were President Theodore Roosevelt and his chief forester, Gifford Pinchot. During the Roosevelt administration (1901–1909), the United States began to pursue a unified national conservation program bent on saving forests, not only for their beauty and shelter for wildlife, but also as a future resource for home construction and employment. Because the program was based on "the use of the natural resources for the greatest good of the greatest number for the longest time,"[21] Pinchot championed practical, professional, and scientific resource management predicated on anthropocentric ethics. He favored management of national forests so as to achieve sustainable yields instead of the common method of cutting the forests all at once for immediate economic gain.

President Roosevelt might be considered the first environmental president. His administration combined an understanding of politics with an idealism about nature and developing an anthropocentric strategy. The springs of Roosevelt's conservation are not wholly clear. Some historians even argue that Roosevelt's principal motive was to maintain the stock of wildlife sought by hunters, because he himself was an avid hunter.

Whatever Roosevelt's motives may have been, while he was president, the Reclamation Act of 1902 led to the establishment of the Bureau of Reclamation in the Department of the Interior, as well as major programs created to reclaim large regions of land in western states. In 1905 Congress transferred 85.6 million acres to a bureau in the Department of Agriculture that would become the U.S. Forest Service. The national forest reserves came under the control of the new U.S. Forest Service, headed by Gifford Pinchot, and 16 million acres of timberland were proclaimed National Forest lands. During Roosevelt's administration, the total federal forest acreage rose from 42 million to 172 million acres. (Refer to Figure 1.1.)

World War I (1914–1918) caused a complete but temporary stop in the conservation movement. During the war, the environmental agenda, if any, "was to exploit the environment as quickly as possible to support the war effort."[22] Nevertheless, the notion of federal responsibility for land management was further fostered in 1916 when President Woodrow Wilson signed the bill that formally established the National Park Service in the Department of the Interior.

Conservation measures again came to the forefront with the advent of the Great Depression (1929) and the severe drought of the 1930s. President Franklin D. Roosevelt initiated a new era in American environmental conservation. In his 1932 acceptance speech to the Democratic Convention, FDR focused national attention on the problems caused by private use of the public domain:

There are tens of millions of acres east of the Mississippi River alone in abandoned farms, cut-over land, now growing up in worthless brush. Why, every European Nation has a definite land policy, and has had one for generations. We have none. Having none, we face a future of soil erosion and timber famine. It is clear that economic foresight and immediate employment march hand in hand in the call for the reforestation of these vast areas.[23]

Franklin Roosevelt's administration promoted both large reclamation projects and large investments of human capital simultaneously to reduce unemployment and spark the economy, and to protect the biosphere. The Civilian Conservation Corps (CCC), established in 1933, had the double purpose of employment relief and conservation. Young men, fifteen to twenty-four years old, who could not find employment were recruited into the CCC to work on forestry, wildlife and park projects, prevention of soil erosion, and flood control, among several other tasks.

The Tennessee Valley Authority (TVA), which had been vetoed during the 1920s, was created in 1933. By building a system of dams in the Tennessee River and its tributaries, the TVA reduced floods and contributed to regulation of the lower Ohio and Missouri Rivers. In the following years additional prime forest lands were selected as "national forests," and regions of scenic or historic importance were added to the National Park System.

Furthermore, the General Land Office resumed the transfer of remaining public lands to private ownership until passage of the Taylor Grazing Act of 1934. This Act more or less terminated the era of major transfer of public lands and created the Grazing Service in the Department of the Interior. This brought control to the unlimited use of public lands by livestock. Rexford Tugwell, one of Roosevelt's group of experts on conservation policy, commented that the Taylor Grazing Act had terminated "private land settlement which has strongly influenced the shaping of American character, and the establishment of individual enterprise as the dominating feature of American economic life."[24] In 1946 the Bureau of Land Management was created to provide a permanent management of the remaining public lands.

The movement to reserve public lands from private ownership and retain them under federal public domain originated from the interaction of both anthropocentric and ecocentric values. Therefore, it is also worthwhile to remember that there is no single polarization between the anthropocentric and ecocentric values. Actually, as pointed out by the geographers Sophia Bowlby and Michelle Lowe:

It was, in fact, people working within the latter tradition [anthropocentric ethics] who were most important in establishing many of the agencies of resource conservation in America. They did so to ensure adequate resources for the future. Moreover, Thoreau himself, and many others with similar ideas, were also concerned to use the earth's resources for the benefit and economic well being of humanity.[25]

An environmental dispute that emerged in 1901 showed that ecocentric values alone were not strong enough to preserve pristine environments. The controversy dealt with a plan to supply water for San Francisco, California, and proposed to build a reservoir in the beautiful Hetch Hetchy Valley of Yosemite National Park. The issue turned on the question of whether an artificial impoundment should be allowed within a national park. Other potential valleys were available, but the Hetch Hetchy Valley was the least expensive.

The dispute polarized the two most influential environmentalists of the time, John Muir and Gifford Pinchot. Pinchot had no sympathy for preservation of pristine environments, viewing them as a waste of society's resources. He championed the reservoir as a utilitarian development—it would benefit the economy of the region. Muir, on the other hand, was an ecocentric who favored the valley in its natural state. To Muir, the reservoir would violate the spirit of national park systems, it would flood a spectacular scenic area, and it would not offer any recreational benefits. Due in part to the politically strong conservationist forces led by Pinchot, the ecocentric values could not stand against the anthropocentric arguments, and in 1913 Congress approved construction of the dam. Congress considered the dam to be a rational use of the region's natural resources.

In later decades, with other similar controversies, environmentalists who opposed the construction of dams because they violated the national park system and jeopardized magnificent scenic areas have resorted to other arguments in order to halt the construction. Proponents of wildlife preservations have utilized studies of their own to support their view that the dams were unnecessary or economically unfeasible. As a result of combined anthropocentric and ecocentric arguments, a number of dams have been dropped from development. These included Echo Park Dam in western Colorado and Bridge Canyon Dam on the Colorado River.

As previously stated, environmental values are prescriptive, thus subject to different views over what constitutes good environmental ethics. Although anthropocentric and ecocentric ethics are normative, they have had, and will continue to have, a great influence upon the development of environmental policy. Witness, for example, the endangered species acts, wildlife refuges, national parks, and other land use controls protecting the intangible qualities of nature. The dwindling biosphere and biodiversity have caused many concerned individuals to reassess humanity's relationship with nature. Some laws have been enacted to control trafficking in endangered species and products such as exotic birds, rhinoceros horns, elephant ivory, reptile leather, and cacti.

In the United States, congressional efforts to protect the biosphere and its species are due primarily to the political actions of environmentalists, ecologists, and the public at large. In response to the political activities of these concerned citizens, Congress has enacted numerous laws addressing concerns as varied as species conservation, the preservation of wild scenic sites, marine protection, forest management, and ocean dumping. An examination of the Endangered Species Act (1973), one of the most far-reaching and controversial of the new federal environmental laws, illustrates both the successes and difficulties in the conservation of biodiversity.

The first federal laws protecting species were clearly predicated on anthropocentric concerns and focused primarily on species deemed important to agriculture and horticulture. The first such law was the Lacey Act, passed in 1900, which prohibited the interstate transportation of certain species, including starlings and sparrows, killed in violation of state laws. The Act authorized the Secretary of Agriculture to take all means necessary for the "preservation . . . and restoration of game birds and other wild birds." By prohibiting interstate transportation, the statute "established the federal government as the enforcer of state game laws."[26] The federal government's second venture in species protection was the Migratory Bird Treaty Act of 1918, by which Congress used its treaty power to protect waterfowls migrating between the United States and Canada.

The Endangered Species Act (ESA) of 1973 was the first formal, federal recognition that each species plays a unique ecological role in the environment and that the government should, whenever possible, avoid disrupting the balance of nature. The dual responsibility for continuous and systematic administration of the ESA rests with the Fish and Wildlife Service (FWS) for terrestrial and some aquatic species and the National Marine Fisheries

Service (NMFS) for marine species and anadromous fisheries. The initial expressed objective of this Act was "to provide a means whereby the ecosystems upon which endangered species and threatened species depend may be conserved." Notwithstanding this clear mandate, most implementation efforts have concentrated on protecting individual species in danger of extinction (endangered) or those likely to become extinct in the foreseeable future (threatened).

The ESA has fundamental problems. At present rates, it would take a few decades to identify species in the United States now considered to be in danger. Moreover, the program has financial limitations. Full implementation of the act would cost a few billion dollars, yet the funding has been about $10 million.

Further, the ESA enters into the picture when options and alternatives are limited. Species identified as threatened or endangered by definition are the ones on the brink of extinction. Often, ecological niches for such species exist only in isolated patches that cannot support ecologically viable populations. Loss of genetic diversity and a lack of self-sustainable natural ecosystems limit captive recovery efforts. From an ecological perspective, a better approach is to concentrate on protecting self-sustainable ecosystems of these species, and letting nature take its course. Evolution by natural selection would favor those species whose genetic traits better adapt them to a particular natural ecosystem. Federal agencies have begun to increase the employment of regional and multispecies habitat-based approaches. To illustrate, FWS emphasizes additions to threatened and endangered species lists that cover multiple species in a given region so that conservation efforts can address their various needs more efficiently. The agency also supports multispecies recovery actions.

Federal agencies have a duty to manage their activities in ways that neither jeopardize those species identified nor destroy or adversely modify their critical environment. The ESA requires that any federal agency

insure that any action it authorizes, funds, or carries out, in the United States or upon the high seas, is not likely to jeopardize the continued existence of any listed species or results in the destruction or adverse modification of critical habitat.[27]

Unfortunately, ESA enters the picture when federal agency options are restricted. Sometimes, courses of action are often mapped out or significant investments have been put into the project before sponsors can determine whether or not there will be an impact on a threatened or endangered species. This problem is illustrated by the famous snail darter controversy, the case of *TVA v. Hill* of 1978.

In *TVA v. Hill*, the Tennessee Valley Authority (TVA), a federally owned corporation, was constructing a dam in the 1960s in order to control flooding, generate electricity, and promote economic growth in the region. People in the region dispossessed by the project, environmentalists, and the Cherokee Indians, for whom the region was a sacred and historic site, were against the construction of the dam. Their opponents were developers and others who wanted to have all federally funded projects exempted from the act.

In a lawsuit brought against the TVA, the Environmental Defense Fund and other environmentalists sought to have construction of the dam enjoined. After the lawsuit began, it was determined that its construction would threaten the survival of a previously unknown perch species named *Percina imostoma tanasi*, popularly known as the snail darter. Nevertheless, Congress continued to finance the construction of this project, which was almost completed by the time the case was heard in federal district court. The district court agreed that building the project violated the Endangered Species Act. Even so, the court declined to enjoin the completion of the dam precisely because of Congress's continued funding of the project and because the project was almost completed. The U.S. Court of Appeals reversed the decision and ruled in favor of the environmental groups. The case was then reviewed by the United States Supreme Court.

In 1978 the Supreme Court ruled that the courts must enforce the laws that the legislature enacted when enforcement was sought. Under the doctrine of separation of powers, once the legislature has exercised its delegated powers and enacted a statute assessing the order of importance of a particular issue, it is the duty of the executive to administer the priorities, and of the court to enforce it when enforcement is sought. It is not the function of the Supreme Court to reassess the wisdom of the order of priorities in a given area decided by the legislature. Once the meaning of a statute is discerned and its constitutionality decided, the judicial inquiry is terminated. In this case, it is quite clear that the Endangered Species Act mandates that completion of the project be stopped in deference to the continued survival of the snail darter. The Court said,

It may seem curious to some that the survival of a relatively small number of three-inch fish among all the countless millions of species extant would require the permanent halting of a virtually completed dam for which Congress has expended more than $100 million. . . . We conclude, however, that the explicit provisions of the Endangered Species Act require precisely that result.[28]

After this landmark case, the Endangered Species Act was amended in 1978 to set up a cabinet-level Endangered Species Committee. This committee was authorized to balance the value of the species against other values, and to grant exemptions in instances where the following criteria are met: absence of reasonable and prudent options; regional or national significance, defined on a case-by-case basis; benefits that outweigh those of alternatives; and identified mitigating measures.

The committee convened and reached a decision that reasonable alternatives were available and that the project was economically unsound. Consequently, it denied an exemption for the TVA dam. In 1979 Congress overruled that decision and exempted the project from the ESA's requirements, and the TVA transferred some snail darters to other streams nearby where the species were surviving.

Another battle over protection of a species is going on today in the northwestern section of the United States. In 1990 the Fish and Wildlife Service identified the northern-spotted owl as a threatened species, and 6.9 million acres of old-growth forests located on federal lands were set aside to ensure the owl's survival. An intense conflict developed between environmentalists, who wanted more acres set aside, and timber-dependent communities, who argued that thirty-three thousand jobs would be lost if the forest were protected from timber harvesting.

In 1993 President Bill Clinton convened a Forest Conference in Portland, Oregon, in order to resolve the old growth–spotted owl controversy. The conference set the stage for a program to ensure sustainable ecosystems and a sustainable economy. After the conference, an interdisciplinary Forest Ecosystem Management Assessment Team formulated ten management alternatives for the region. From these options, Clinton selected the watershed-based plan. This plan included three key provisions.

First, a planning and monitoring program would be established in order to manage the old-growth ecosystems in the public lands concerned. Second, a complex of old-growth reserves, riparian reserves, Adaptive Management Areas, and forest management matrix would be established across the twenty-five million acres of the region. Third, the plan provides for sustainable annual sales of 1.1 billion board feet of timber. Clinton hoped that by adopting these provisions, a satisfactory compromise had been reached.

Apparently, judging by the results of the snail darter and spotted owl controversies as well as other cases, "There is a growing wave of opinion, however, that the debate over the future of endangered species (and biodiversity in general) is becoming more political and less scientific in nature."[29] Moreover, in present-day America, conservationists' policies still prevail.

During President Clinton's delivery of the 1993 Earth Day Address, one could hear an echo of the anthropocentric ethics:

Our forest plan is a balanced and comprehensive program to put people back to work and protect ancient forests for future generations. It will not solve all of the region's problems but is a strong first step at restoring both the long-term health of the region's ecosystem and the region's economy.[30]

Global Programs Concerning the Biosphere

Migratory wild animals do not recognize mankind's political boundaries. They migrate from one nation to another as well as through international zones. Animal migrations are dictated by a need to seek warmer weather as well as a need for favorable locations for feeding and/or mating. Examples of migratory animals are monarch butterflies, certain crustaceans and squids, many fish, green turtles, antelopes, wildebeests, some whales and dolphins, many bats, emperor and Adelie penguins, arctic terns, and robins.

Worldwide efforts to protect these international travelers and their ecosystems are being made under the auspices of many governmental and institutional organizations such as the World Wildlife Fund, Nature Conservancy, National Audubon Society, Sierra Club, and Environmental Defense Fund. (See the Primary Documents chapter for lists of major environmental organizations involved with environmental issues.) These public and private organizations have a variety of options to explore in their efforts to preserve prime natural biomes. Efforts to save biodiversity include private purchase of pristine environments for preservation, and the so-called debt-for-nature swaps, whereby developing nations receive money if they protect ecologically sensitive ecosystems. Natural area zoning can be used to protect wetlands, coastal dunes, wild rivers, game preserves and other natural habitats, forests, fish spawning areas, and other vital areas.

Other current biome-conservation issues of international concern include: (1) conservation of large representative ecosystem reserves and natural sites of world significance; (2) maintenance of regional wildlife management and conservation training in institutions, such as the College of African Wildlife in Tanzania; (3) research in the minimal critical size of ecosystems required for viable wildlife, which is currently taking place in Brazil; and (4) the U.N. Convention on Biological Diversity.

To date, the U.N. Convention on Biological Diversity (1992) has been the most comprehensive international effort to conserve biodiversity. The planning for the convention dates back to 1987 when the United Nations

Environment Programme (UNEP) first called attention to the need for international laws protecting Earth's biodiversity. In 1988 an Ad Hoc Working Group of Experts on Biological Diversity was established to consider the most-effective means of protecting biodiversity. Based on the group's recommendation, UNEP created the Negotiating Committee for a Convention on Biological Diversity in order to negotiate an agreement that all nations could agree upon.

In 1992, after two years of concentrated negotiations, 153 nations signed a treaty that mandates measures to conserve biodiversity through programs that encourage countries to, first, adopt sound domestic conservation programs, such as protecting their species with laws like the U.S. Endangered Species Act, creating national parks, and protecting regions. Second, the convention, in order to encourage sustainable economic development, must promote the coupling of commerce to biodiversity conservation. Third, the convention must recognize that much of Earth's biodiversity lies in developing nations, and that benefits deriving from biotechnology must flow back to those nations acting to conserve biodiversity. According to the Council on Environmental Quality:

These benefits—determined on the basis of voluntary agreements among all concerned—could take the form of monetary compensation for the use of genetic resources, or as technology transfer programs in training, participation in research, cooperative work programs, and improved access to information.[31]

Fourth, the biodiversity convention establishes an international forum for nations to share their experience and knowledge on the conservation and sustainable use of wildlife. Participation in the forum should be helpful in building a strategy for implementing the goals of the convention.

In effect, the convention requires nations to identify biological resources in decline, the reasons for such declines, and on- and off-site measures to address the decline. Significantly weakening the force of the convention, however, was the fact that, although the United States participated intensively in the negotiations, President George Bush did not sign the treaty. Among the reasons given by the United States was that the convention included provisions unrelated to the conservation sections that were unacceptable. These provisions dealt with intellectual property rights, access to genetic information used in biotechnology, and international funding for biodiversity activities.

One year later, the new U.S. president, Bill Clinton, signed the treaty, emphasizing the importance of conservation and sustainable use of biodiversity worldwide. Presently, it is too early to tell whether or not this forum will prove to be an effective mechanism for protecting the world's rich biological inheritance for future generations. As one observer has noted, "The convention suffers, however, from a lack of targets, timetables, and enforcement mechanisms."[32]

In conclusion, the emergence of the first wave of environmentalism involves remarkable changes in private and public policies regarding nature. Predicated on the anthropocentric and ecocentric ethics, environmentalists are causing society to recognize the value of the biosphere and its genetic diversity. However, achieving these policies presents many difficult problems. Scientists do not know much about the ecological niches of most species. The gaps in scientific knowledge are most severe with respect to the understanding of how the various ecosystems interact to form the ecosphere. Furthermore, the ecospheric approach requires interdisciplinary, interjurisdictional, holistic, and adaptive efforts that are new to private and public decision makers.

Concerned efforts are being made to conquer these challenges. What is most promising is that conservation efforts are increasingly based on cooperation, negotiation, and partnerships among decision makers. Also, a growing number of statutes and treaties have been enacted and ratified to protect ecosystems and their genetic diversity. In the face of increasing population growth and development, ultimately conservation means changing our focus from how we can parasitize nature to how we can live mutualistically with nature.

NOTES

1. Sandra Postel, "Halting Land Degradation," in Lester R. Brown et al., eds., *State of the World* (New York: W.W. Norton & Co., 1989), 23.

2. Don Paarlberg, *Toward a Well-Fed World* (Ames: Iowa State University Press, 1988), xv.

3. Andrew Goudie, *The Human Impact on the Natural Environment* (Cambridge, MA: The MIT Press, 1986), 37.

4. Food and Agriculture Organization of the United Nations, *Forest Resources Assessment, 1990: Tropical Countries* (Rome: FAO, 1993).

5. United Nations Environment Programme, *The State of the Environment* (Nairobi: UNEP, 1977).

6. Jean E. Gorse and David R. Steeds, *Desertification in the Sahelian and Sudanian Zones of West Africa* (Washington, DC: World Bank, 1987).

7. Edward M. Anson, "The Ecology of the Ancient World," in Lester J. Bilsky, ed., *Historical Ecology* (Port Washington: Kennikat Press, 1980), 44.

8. Carolus Linnaeus, *Discovery of Nature*, 1749.

9. Lynn White, Jr., "The Historical Roots of Our Ecologic Crisis," *Science* 155 (1967): 1207.

10. Genesis 1: 26–27, King James Version.

11. Aldo Leopold, *A Sand County Almanac* (New York: Oxford University Press, 1949), 176–77.

12. Bradford Torrey and Francis H. Allen, eds., *The Journal of Henry David Thoreau* (Boston, MA: Houghton Mifflin, 1906), 306–7.

13. Cited in Tom H. Watkins, "Why a Biographer Looks to Muir," *Sierra Club Bulletin*, May 1976, p. 18.

14. Jimmie Durham, *Columbus Day* (Minneapolis, MN: West End Press, 1983), 70.

15. Cited in Encyclopedia Britannica, *Macropaedia* 4: 1092, 1975.

16. Robert H. Nelson, "Unoriginal Sin: The Judeo-Christian Roots of Ecotheology," *Policy Review* (Summer 1990): 52.

17. Leopold, note 11, p. 240.

18. Constitution of the Ecological Society of America, Article 15 (c & d).

19. Mark Sagoff, "Fact and Value in Ecological Science," *Environmental Ethics* 7 (1985): 99–100.

20. George P. Marsh, *Man and Nature* (Cambridge, MA: Harvard University Press, 1864), 42, 203.

21. Cited in Stewart L. Udall, *The Quiet Crisis* (New York: Holt, Rinehart and Winston, 1963), 71–72.

22. Melvin G. Marcus, "Environmental Policies in the United States," in Chris C. Park, ed., *Environmental Policies* (London: Croom Helm, 1986), 49.

23. Aaron Singer, *Campaign Speeches of American Presidential Candidates 1928–1972* (New York: Frederick Ungar Publishing Co., 1972), 73.

24. Rexford G. Tugwell, "Our New National Domain," *Scribner's Magazine* XCIX (March 1936): 168.

25. Sophie R. Bowlby and Michelle S. Lowe, "Environmental and Green Movements," in Antoinette M. Mannion and Sophie R. Bowlby, eds., *Environmental Issues in the 1990s* (England: John Wiley & Sons, 1992), 164.

26. Editorial Research Report, *Environmental Issues: Prospects and Problems* (Washington, DC: Congressional Quarterly, Inc., 1982), 126.

27. 50 Code of Federal Regulation, part 402.01(a).

28. *Tennessee Valley Authority v. Hill*, 437 US 153 (1978).

29. Jacqueline V. Switzer, *Environmental Politics: Domestic and Global Dimensions* (New York: St. Martin's Press, 1994), 315.

30. Cited in Council on Environmental Quality, *Twenty-Fourth Annual Report* (Washington, DC: Superintendent of Documents, 1993), ix.

31. Ibid., 233.

32. Christopher Flavin, "The Legacy of Rio," in Lester R. Brown et al., eds., *State of the World 1997* (New York: W.W. Norton, 1997), 14.

3

Pollution and the Emergence of Environmentalism

Many environmental historians consider the development of new technologies and their polluting by-products to be the root cause of the second wave of environmental concern. This phase of rampant pollution has sparked many antipollution movements.

As described in Chapter 2, the concern for the vanishing wilderness largely attracted people from the middle and upper classes to the environmental cause. Key environmentalists were concerned individuals, such as John Muir and Aldo Leopold. They were citizens from the higher rungs of the social ladder who pondered the values of nature and the impact that its disappearance would have on society. These early environmentalists were highly influential and garnered support from the government in their effort to preserve the pristine environment. The antipollution movement, over time, has attracted a greater number of advocates from diverse socioeconomic backgrounds, primarily because people of all social and economic classes can see and feel the effects of pollutants. In fact, however, pollution hits the poor the hardest: Low-income residents of dilapidated buildings and migrant farm-workers are more susceptible to lead poisoning in their homes or are disproportionately likely to acquire pesticide poisoning.

This chapter focuses on how pollutants have stirred society's awareness of its environment and on the principal players who brought the pollution issues to the forefront, and it considers the problems and prospects of regulating pollutants. In order to develop these issues, it is necessary to describe some relevant aspects of pollutants.

RELEVANT ASPECTS OF POLLUTANTS

In general, pollution or habitat contamination occurs because society lacks some mature, well-developed methods of recycling or reusing resources that would otherwise be disposed of as waste. One type of pollution is in reality nonrecoverable matter, or waste heat. In fact, the term *waste heat*, or thermal pollution, is used to describe a human-induced alteration of natural water temperature. Steam electric-power plants are the most common sources of thermal pollution. These plants discharge large volumes of warm water that have been used as cooling agents in the process of generating electricity. This type of pollution can have disastrous consequences on aquatic organisms in the nearby ecosystem. In general, increased temperature decreases the level of dissolved oxygen available to aquatic organisms. Certain species (e.g., largemouth bass, catfish, crappies, sunfish) prefer water that contains less oxygen, whereas trout prefer water that contains a large supply of oxygen. Consequently, a change in temperature directly affects the oxygen concentration and indirectly affects species composition. Heated water also may increase the toxic effects of certain pollutants. Furthermore, most organisms are able to exist only within a certain temperature range. Therefore, increased temperature both directly and indirectly alters the species composition of aquatic systems.

Pollution is one of the most pervasive environmental concerns. The growth of industries and cities placed a tremendous burden on the environment by introducing pollutants such as pesticides, radioactive isotopes, and heavy metals into the air, land, and water. There are now approximately seventy thousand different chemicals in the marketplace, and new ones are being produced at an accelerating rate. In addition, the use of precious supplies of energy and materials for luxuries such as motorboats, air conditioners, and hair dryers has added more stress to the environment. These items deplete limited resources of raw materials necessary to make such products, as well as the depletion of "clean air" in the environment. The use of natural resources at a rate higher than nature's capacity to restore itself will result in the pollution of air, water, and land.

From the standpoint of the environmental crisis, it is necessary to recognize three general characteristics of pollutants:

1. Pollutants recognize no boundaries.
2. Many toxins cannot be degraded by organisms and consequently persist in the ecosphere for many years.
3. Pollutants destroy biota and habitat.

Pollutants Recognize No Political Boundaries

Pollutants such as waste heat, chlorofluorocarbons (CFCs), radiation, pesticides, carbon dioxide, and other gases are found in air, land, or water. They are capable of moving from one sphere to another. Usually, chemicals on the land are carried by rain water to nearby waterways where they may cause pollution. The activities of one individual can create pollution that is detrimental to other individuals or to the society at large.

The transboundary movements of pollutants lead to unusual and complex economic and political difficulties. Individuals in one nation may suffer economic loss and health hazards as a result of the pollution originated in another nation, yet they may not benefit from the economic activity that caused the pollution. On the other hand, a nation or jurisdiction that acts to minimize pollution that is being transported over hundreds of miles may gain little local environmental benefit. The interjurisdictional problems related to the regulation of long-range air pollution are especially apparent in the acid rain issue, which has led to several international disagreements. For example, a large amount of the acid precipitation that is damaging the environment of the northeastern United States and eastern Canada is caused by industrial activity in the American Midwest.

The solution to transboundary pollutants is unlikely to be found within a nation's borders. It is increasingly becoming apparent that the transport of pollutants over long distances raises the question of global pollution such as ozone depletion, acid rain, and the greenhouse effect. The discovery of the transboundary nature of pollutants is forging a worldwide consensus that international laws are necessary to protect the global commons.

Pollutants May Persist in the Ecosphere

The processes responsible for the alteration or disintegration of pollutants may, for clarification, be labeled as biological and physicochemical. The decomposition of pollutants through the biological activities of organisms, known as biological disintegration, has long been cited as the cause of disintegration of pollutants. Scientists identified two classes of pollutants in relation to their decomposition by living organisms: biodegradable and nonbiodegradable. Biodegradable wastes are capable of being broken down or used by organisms. They include organic waste products, phosphates, and inorganic salts. Nonbiodegradable pollutants resist decomposition by organisms and remain in the environment for an indefinite period of time. Nonbiodegradable pollutants include bottles, cans, metals, plastics, certain pesticides and herbicides, and radioactive isotopes.

Biodegradable pollutants are temporary nuisances that organisms break down into harmless compounds. If the pollutant is organic (including carbohydrates, proteins, fats, and nucleic acids), the organism obtains energy and/or material for its own use in the process of breaking it down. However, biodegradable pollutants could have serious environmental consequences if large quantities are released in a small area. For example, the dumping of organic or food waste into a small pond will deplete the pond's oxygen supply. Left with no oxygen, the fish will die. Thus, biodegradable substances become pollutants when they overload the environment because they cannot be broken down by organisms at a rate fast enough to maintain the integrity of the environment. Ecologists use the term "assimilative capacity" to express the ability of an aquatic ecosystem to assimilate a substance without degrading or damaging its integrity. Integrity is generally defined as maintenance of the structure and the functional characteristics of a system.

Nonbiodegradable pollutants, on the other hand, are dangerous simply because organisms have neither evolved enzymes capable of digesting them, nor have they developed a defense mechanism against them. Fat-soluble nonbiodegradable pollutants, such as methylated mercury, chlorinated hydrocarbons (for example, DDT and PCB), benzene, and polyaromatic hydrocarbons, have an additional and more significant property. Because they are fat soluble, but not water soluble, these pollutants are not excreted in urine. Instead, they accumulate in the fat of organisms. Because organisms cannot metabolize these toxins, they retain almost 100 percent of them. Therefore, toxins can be concentrated within organisms far beyond their environmental levels (see Figure 3.1). This process is called biomagnification. The problem is serious in the higher levels of the food chain.

PCBs are chemicals that were widely used in electrical equipment before they were banned in 1976 for being toxic to humans. During biomagnification, chemicals are retained in fatty body tissues and increase in concentration over time. The chemicals bioaccumulate in species higher in the food chain as contaminated food species are eaten. Figure 3.1 illustrates how the PCBs are almost 50,000 times as concentrated in herring gull eggs as they are in phytoplankton. Scientific evidence indicates that pregnant women who regularly ate fish contaminated with PCBs gave birth to a significantly higher number of babies with low birth weight who later suffered delayed intellectual and physical development as children.

The transfer of energy from lower to higher food levels is extremely inefficient. Biomagnification results from the increasing concentration of toxic substances within each successive link in the food chain. For instance, herbivores must eat large quantities of plant materials (which may be contami-

Figure 3.1
Biomagnification of Polychlorinated Biphenyls (PCBs) in the Aquatic Food Chain of the Great Lakes

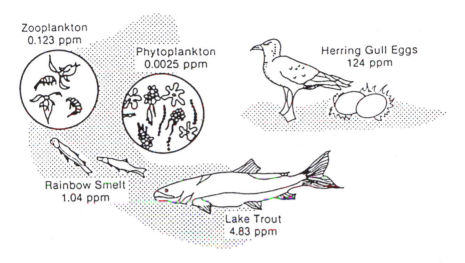

Zooplankton
0.123 ppm

Phytoplankton
0.0025 ppm

Herring Gull Eggs
124 ppm

Rainbow Smelt
1.04 ppm

Lake Trout
4.83 ppm

Source: U.S. Environmental Protection Agency and Environment Canada. *The Great Lakes: An Environmental Atlas and Resource Book.* Chicago and Toronto, 1987.

nated with toxins), carnivores must eat many more herbivores, and so on. Because fat-soluble toxins are not excreted, the carnivore gets accumulated dosages of toxins from its food over a long period of time. Thus, in the higher trophic levels of carnivores, including humans, these toxins have reached high concentrations. As a result of biomagnification in the food that they eat, many people have concentrated pollutants in their bodies. For example, in the United States dichlorodiphenyltrichloroethane (DDT) has been found in human breast milk as high as 0.5 parts per million. This is 100 times the Food and Drug Administration's maximum amount of DDT permissible in human food. However, the long-term effects of such low concentrations in humans remain unknown.

Nonbiological disintegration of pollutants is complex. Several factors, such as wind, water, and climate, generally act together to achieve decomposition or decay. There seems to be very little doubt that a pollutant like carbon monoxide reacts strongly with oxygen in the atmosphere and is rapidly converted to carbon dioxide. Moreover, radioactive isotopes by themselves will eventually decay into harmless substances. The latter process may take days or thousands of years, depending on the isotope.

During the past few decades, scientists have become increasingly aware that some pollutants may be decomposing more slowly than expected. According to the Council on Environmental Quality:

One important aspect of the ozone and global warming issues is that the atmospheric lifetimes of gases such as nitrous oxide, and Chlorofluorocarbons 11 and 12, are known to be very long (as much as 180 years). Consequently, if there is a change in atmospheric ozone or climate caused by increasing atmospheric concentrations of these gases the full recovery of the system will take several tens to hundreds of years after the emission of these gases into the atmosphere is terminated.[1]

Toxic Pollutants

Based upon what toxic pollutants affect, we can group them into two categories: biocides (from *bio*[life] and *cide*[kill]), and locucides (from locus [place]). Biocidic pollutants harm organisms directly by affecting specific tissues or entire organisms. From the ecological point of view, locucidic pollutants are more dangerous because they have an adverse effect on the natural environment or habitats of species.

When a pollutant reduces or denatures (changes) a locus, that pollutant can be classified as locucidic. Obviously, as more space is taken up by people, buildings, and domesticated organisms, less habitat (locus) remains available for other species. Locucidic pollutants have the potential of affecting the ecosphere in both dramatic and uncertain ways. The global ecosphere is being influenced by locucidic pollutants such as discharges of chemicals (CFCs, carbon dioxide, and other gases), deforestation, and other habitat modification. As described at the end of this chapter, both the greenhouse warming and the possibility of ozone depletion are examples of ways in which locucidic pollutants are changing the ecosphere. Enormous uncertainties for scientific measurement exist because various parts of the ecosphere interact in innumerable ways on scales ranging from molecular to global. Locucidic pollutants are changing the ecosphere on a grand scale, with the outcome still in question, and the consequences potentially far-reaching.

Biocides can harm specific parts of an organism (organ specific) or an entire organism (systemic toxicity). For example, too much DDT in a bird's tissue interferes with the deposition of calcium in its eggshell. As a result, birds lay eggs with shells that are much thinner than normal. These fragile eggs break easily, and provide no protection for development inside the egg. Predatory birds are the ones most seriously affected because they occupy the higher trophic levels.

Most pollutants denature both the locus and species simultaneously. For example, a prolonged increase or decrease of temperature beyond the tolerance range of species denatures the locus while simultaneously affecting the organism's physiology. Consequently, the prolonged changes in temperature may kill the organism directly by changing its physiology, or indirectly by denaturing its locus.

Another important characteristic of pollutants is the level of exposure required to cause damage. The effects of pollutants can be immediate (acute toxicity). Here, an affected organism experiences the toxic effect from short-term exposure to a pollutant. Possible life-threatening effects are neurological damage; damage to the respiratory system, liver, or kidney; damage to the immune function; and acute teratogenic effects (congenitally malformed babies). On the other hand, high-level exposures to some pollutants for a short period of time may produce acute, though often temporary effects, such as rashes, burns, or poisoning. The effects of a pollutant can also be long-term (chronic toxicity). The toxic effect is produced by exposure of a continuing and prolonged nature. Of concern are delayed toxic reactions, progressive degenerative tissue damage, reproductive toxicity, cancer, mutations, and various other problems. It has been estimated that about 80 percent of human cancer is induced by carcinogenic substances found in the consumption of food, water, and air.

The range of risks depends upon the extent of exposure of the population (how much exposure to different levels of a pollutant), and the potency of the pollutant. For example, a highly toxic pollutant is risk-free if living creatures are never exposed to it.

Although each pollutant in the environment may not be present in amounts that cause environmental damage, it is imperative to note that the interaction of pollutants is inevitable. The additive effects of many toxic pollutants in low concentrations may place significant stress on the global ecosphere. Each pollutant increases the range of factors to which organisms must respond. Synergistic effects (the effects of a combination of pollutants) may exert a greater influence than the sum of their individual effects. The synergy between asbestos exposure and smoking in causing lung cancer is a well-known example of a synergistic effect. On the other hand, some toxic substances may have antagonistic effects that could reduce the overall toxicity of the added effects. As emphasized earlier, it is easier to deal with individual pollutants in isolation. As will be described at the end of this chapter, the chemicals CFCs, carbon dioxide, methane, and other gases, which are expected to deplete the ozone layer, are the same chemicals that are expected to warm the ecosphere. Moreover, increasing concentrations of methane are also expected to increase ozone in the atmosphere, and at the same time may contribute to forest damage, which is already occurring.

POLLUTANTS AND "SILENT SPRING"

In the preceding section we reviewed some general characteristics of pollutants. During the 1960s, much of the information described was available in scholarly publications and textbooks. But because this information was not readily available and hard to interpret, overall public awareness was minimal. One event that informed society of the deadly nature of pollutants, and helped make "ecology" and "environment" household words, was the publication in 1962 of *Silent Spring* by Rachel Carson. Because the book was written for general audiences, and was easy to understand, it became a landmark best-seller. Carson was a member of the rare breed of scientists who are able to translate the technical terms of their fields into everyday language.

Silent Spring clearly demonstrated to many individuals that pollution is more than an eyesore—it threatens the complex processes of life itself. Carson's book provided an ecological perspective to the environmental movement. It dramatized how exposure to toxic substances could cause cancer, and how pesticides like DDT affected wildlife and humans. In the book, she stressed the interrelatedness of all species; the dependence of human life, health, and welfare on natural processes; and the warning of a "silent spring" when no birds and insects are left to provide the sounds of spring. In her clear and honest prose she warned:

We stand now where two roads diverge. But unlike the roads in Robert Frost's familiar poem, they are not equally fair. The road we have long been traveling is deceptively easy, a smooth superhighway on which we progress with great speed, but at its end lies disaster. The other fork in the road—the one "less traveled by"—offers our last, our only chance to reach a destination that assures the preservation of our earth.[2]

Many environmental historians consider her book to be the single most important event in launching the second wave of the environmental movement. As Robert Mitchell, a professor of geography, observed:

Carson's success as the environmentalist Paul Revere in awakening people to the potential threats posed by pesticides was a significant achievement, because these second-generation problems are inherently more difficult to communicate to the public than the threats posed to wildlands and "critters" by development projects.[3]

Another influential book that propelled the early stages of the environmental movement was Barry Commoner's *The Closing Circle* (1971). In his book, Commoner blamed the environmental crisis on modern technologies that tend to replace older, less-polluting technologies. Commoner, a biologist, studied the growth rates of a large number of economic processes and placed the blame for the present condition of the environment on post–World War II developments in technology. According to him, technology has promoted a variety of products and activities whose environmental effects are singularly detrimental. These new products and activities replaced older ones whose effects on the environment were relatively benign. In general, new technologies rely heavily on artificial energy sources and generate large quantities of nonbiodegradable wastes. Commoner examined aluminum and showed that its production required six times more energy and generated several times more pollution than the production of steel. In spite of these drawbacks, society prefers aluminum because it has numerous advantages for packaging and construction and for its resistance to rust.

The Closing Circle further rekindled the ecological worldview into the environmental movement. Commoner points out:

The amount of stress which an ecosystem can absorb before it is driven to collapse is also a result of its various interconnections and their relative speeds of response. The more complex the ecosystem, the more successfully it can resist a stress. . . . Like a net, in which each knot is connected to others by several strands, such a fabric can resist collapse better than a simple, unbranched circle of threads—which if cut anywhere breaks down as a whole. Environmental pollution is often a sign that ecological links have been cut and that the ecosystem has been artificially simplified and made more vulnerable to stress and final collapse.[4]

Through the publications of *Silent Spring* and *The Closing Circle*, and other events, Americans became aware of the danger of the indiscriminate use of pesticides. Public concern regarding the use of pesticides and other chemicals led to comprehensive federal regulation of chemical products and chemical wastes. (The Primary Documents chapter contains excerpts from the most important federal environmental statutes that address various pollution problems: water, air, land, and multimedia.)

TRANSBOUNDARY POLLUTANTS AND INTERNATIONAL LAW

Human systems are open systems for matter and energy. Countries are identified by their human communities and political boundaries. Even more so, they are molded by their surroundings. Their food production is coupled with the climate and weather, and their people are coupled with the kinds of foods provided by their agriculture. Plant productivity is affected by light intensity, carbon dioxide availability, temperature, mineral nutrients, and other factors. Some of these factors come largely from external surroundings: light and heat from the sun, oxygen and rainwater from the global commons, and raw materials and manufactured products from imports.

One of the root causes of environmental problems is that the majority of resources in their natural state, such as the atmosphere and the open ocean, are "free goods" that by definition have a price tag of zero. That is, they are resources that do not belong to any specific individual but belong freely to the whole community. As a result, no specific individual or jurisdiction has any incentive to restrict the depletion or avoid polluting these free goods because he or she does not have the right to realize a monetary return from doing so. Consequently, when determining prices of goods and services, prices of the raw materials are not fully considered in the budget of individuals and businesses. In calculating the cost of transacting business, the additional expenses incurred by persons as a result of living in polluted environments are not taken into account. They are therefore outside (external) of the direct cost of manufacturing. Economists call them "externalities."

The biologist Garrett Hardin, in his well-known 1968 essay on "The Tragedy of the Commons," was one of the first individuals to point out that a major cause of resource depletion is the assumption that air and water are free goods. According to Hardin, where there is common ownership of a resource, no one person assumes responsibility for husbanding it. Hardin used the example of cattle grazing on a public pasture:

Picture a pasture open to all. It is to be expected that each herdsman will try to keep as many cattle as possible on the commons. . . . As a rational being, each herdsman seeks to maximize gain. . . . The rational herdsman concludes that the only sensible course for him to pursue is to add another animal to his herd. And another; and another. . . . But this is the conclusion reached by each and every rational herdsman sharing a commons. Therein is the tragedy. Each man is locked into a system that compels him to increase his herd without limit—in a world that is limited.[5]

If every nation, jurisdiction, or individual in the world comes to this same conclusion, there is no hope for the global commons. When a jurisdiction consumes resources without considering the interjurisdictional environmental damage, it disperses this negative attitude throughout the whole community. Yet, when this jurisdiction consumes the resources and manufactures a product, this positive component is a gain for itself alone. Individual jurisdictions are encouraged to minimize their costs irrespective of the environmental damage that may result. When the principle of economics operates, the damage that is done to both the resources and the global commons is not calculated into the cost-benefit analysis.

Municipalities are often able to pass costs onto neighboring communities by dumping sewage into waterways or by building landfills or wastewater treatment plants at municipal borderlines. Local politicians strive to keep taxes down and win re-election by externalizing the costs of their own municipality. National governments also have done their share of externalizing. Many nations have lowered pollution standards in order to encourage industrial development. Countries such as Kuwait and Bosnia have reduced standards because they are fighting for their own national survival. Countries—as do individuals, firms, and municipalities—consider the environmental services provided by the ecosphere as free commodities.

As already noted, the transboundary issue has been politically divisive due to socioeconomic costs and benefits that have been accrued to different nations. The root of this problem is the "Tragedy of the Commons." Consequently, one of the most important adjustments to be made in the operation of the international economy is the internalization of the ecospheric cost when doing cost-benefit analysis. This fact underscores the need for international cooperation in the management of the global commons. Nevertheless, no comprehensive international legal agreement exists to control transboundary pollutants.

Table 4 in the Primary Documents provides a brief overview of the most important international agreements that form the basis of current international law. Space constraints prevent a discussion of each of these treaties. The following two sections focus on the underlying trends, institutional shortcomings, and policy dilemmas that government and nongovernment actions face in attempting to resolve two key international issues—ozone depletion and global warming.

The Concern for the Ozone Layer

The ozone problem is one of the Earth's most pervasive environmental concerns. During the past decade or so, scientists and public decision makers have become increasingly interested in analyzing the processes that control atmospheric ozone. This is due to scientific warnings that human activities may have inadvertently and irreversibly depleted the ozone layer in the upper atmosphere. As far as we know, ozone depletion seems to be linked to a combination of meteorological factors and ozone-depleting air pollutants (i.e., chlorofluorocarbons, or CFCs). This phenomenon was first observed in 1985 when an "ozone hole" appeared in the stratosphere over Antarctica. In 1988, scientists reported that the global ozone layer may be declining as well.

In 1995 for the first time a Nobel Prize was awarded to environmental-related research. The scientists Sherwood Rowland, Mario Molina, and Paul Crutzen were given the Nobel Prize in chemistry for their discovery in the 1970s that CFCs could deplete the stratospheric ozone layer. Their pioneering research helped spark concern for the deteriorating global commons. This concern led to the Montreal Protocol, which was a treaty devised to protect the ozone layer. In order to understand the problem of the depleting ozone shield and the importance of this treaty, the danger of ultraviolet radiation should first be discussed.

The sun emits a large amount of ultraviolet radiation, but the Earth's ozone layer in the upper atmosphere (stratosphere) shields the planet from this kind of solar radiation. This is fortunate because UV radiation is absorbed by nucleic acids (genetic information is stored in nucleic acids). The effects of UV radiation involve the excitation of molecules such as nucleic acids, which then form cross-links. These distortions of nucleic acids interfere with protein synthesis and can cause mutations and cancer. Consequently, many scientists have predicted that various life-forms will be damaged by excess radiation.

The epidermis, which is the outer skin that covers the more sensitive inner tissue, is what protects organisms from excess radiation. Moreover, in some species such as humans, specialized cells in the skin produce melanin. Melanin is the pigment responsible for skin color. Light-skinned people have cells that produce smaller quantities of melanin than do the cells of dark-skinned people. This characteristic is inherited. The production of melanin, however, can be increased by exposure to the sun, resulting in the bronze coloring of the skin commonly known as a suntan. In this process, the pigment absorbs much of the UV radiation and thereby protects the skin from further injury. However, if the skin is overexposed to the sun, the pigment cannot absorb the excess UV radiation, and the skin is injured. The skin becomes inflamed and over the years gets wrinkled. In addition, the radiation can cause dividing cells to become cancerous. Increased exposure to UV also has been associated with the increase of cataracts and the suppression of the human immune system.

As the ozone layer diminishes, the intensity of biologically damaging, ultraviolet radiation in natural daylight increases. A 16-percent ozone decrease will produce an increase of about 44 percent at mid latitudes, and a

30-percent ozone decrease would increase radiation approximately 100 percent. The National Academy of Sciences predicts that a 16-percent ozone depletion would eventually cause several thousand more cases of melanoma per year in the United States. Many of these cases would be fatal, and thousands of other cases of skin cancers would arise. A 16- to 30-percent depletion of ozone is also likely to reduce crop yields of several kinds of plants. Larval forms of several important seafood species, as well as microorganisms at the base of the marine food chain, would suffer an appreciable killing as a result of this 16- to 30-percent ozone depletion.

The "ozone hole" issue has been difficult to resolve because relevant factors (i.e., ozone depletion, acid deposition, and greenhouse warming), to some degree, have been associated with changes in atmospheric composition. For several decades, scientists sought to understand the complex interplay among the chemical, radiative, and dynamical processes that govern how ozone is formed in the atmosphere.

Oxygen, in addition to forming the stable molecule O_2 also can exist in another exceedingly reactive molecular form called ozone (O_3). The characteristic odor that can often be detected around certain electrical equipment is caused by ozone. Even in very small concentrations, ozone can easily be recognized by its distinctive odor. In sufficient quantities, ozone damages lung tissue, reduces lung function, and sensitizes the lungs to other irritants. It also can damage forests and crops.

Oxygen also is formed when nitrogen dioxide (NO_2) absorbs ultraviolet radiation and splits into nitric oxide (NO) and one oxygen atom (O); in a subsequent reaction O combines with O_2 to form O_3. Current studies suggest that ozone formed at ground level by this mechanism is one of the major constituents of photochemical smog. Another method that forms ozone in limited quantities in the upper atmosphere is direct absorption of ultraviolet radiation by oxygen. Oxygen splits into two oxygen atoms ($O_2 \longrightarrow O + O$) that subsequently combines with O_2 to form ozone ($O + O_2 \longrightarrow O_3$). Ozone produced by this natural mechanism forms a protective layer in the stratosphere, shielding the Earth against the sun's harmful ultraviolet radiation.

Numerous scientists are studying the effects of the continuing release of CFCs into the atmosphere. CFCs deteriorate Earth's protective layer of ozone. CFCs are a class of chemical compounds used as solvents, aerosol propellants, blowing agents in foam, and refrigerants. CFCs deplete ozone in the stratosphere, and as a result, increase absorption of damaging UV radiation. (The chlorine released from CFCs is believed to reduce stratospheric ozone concentrations; one molecule of a CFC destroys thousands of molecules of ozone.)

To prevent ozone layer depletion and eliminate one category of greenhouse gases, the United States, Canada, Sweden, and Norway banned nonessential use of CFCs in aerosol spray cans in the 1970s. Other countries, such as Austria, Australia, Japan, New Zealand, and Switzerland, also have established controls over, or have voluntarily limited, production of CFCs. The result of ongoing worldwide research efforts later set the stage for an international banning of CFCs.

International organizations like the United Nations Environment Program (UNEP), concerned with ozone depletion, recognize in particular that effective control needs coordinated global ratification. In 1982 delegates from about twenty nations met in Sweden under UNEP auspices to begin a discussion on the requirement of an international framework for the protection of the ozone shield that would

(1) harmonize regulatory control actions on ozone modifying substances at the international level, (2) increase coordination of ozone related research, and (3) increase the exchange of information on all scientific, technical and legal issues relevant to the ozone issue.[6]

In 1987 the United States and twenty-three other countries agreed to the Montreal Protocol, which regulates the use of CFCs. According to this treaty, the production of CFCs will be cut to half its 1986 level by 1995.

Studies of UNEP and others indicate that ozone depletion is now observed year-round at mid-latitudes in both hemispheres, with winter and springtime declines of as much as 6 to 8 percent observed poleward of 45 degrees. Prompted by this fact, the Protocol was amended in 1990 in London and again in 1992 in Copenhagen. In the subsequent London amendment, members of the Protocol established a three-year fund to aid developing countries to undergo the transition to CFC-like substitutes that decompose more quickly in the atmosphere. Two new ozone-depleting compounds—carbon tetrachloride and methyl chloroform—were added to the list of controlled chemicals. Parties to the treaty and the amendment agreed to phase out the production of CFCs and other ozone-depleting chemicals by the year 2000. The 1992 Copenhagen amendment accelerated the complete phaseout date of CFCs to January 1, 1996 (Table 3.1). By 1993 over 140 countries had ratified the Montreal Protocol.

An important feature of the Montreal Protocol is its technology-forcing aspect that requires sources of chemicals that deplete the ozone layer to meet a schedule of compliance with air pollution standards. Consequently,

Table 3.1
Chemicals That Destroy the Stratospheric Ozone Layer

Chemical	Use	Lifetime in Atmosphere (Years)	Cease Date for Production and Import
Chlorofluorocarbons (CFCs)	Refrigerators, car air conditioners, foam cushioning & insulation, solvents, sterilants	100	January 1, 1996*
Halons	Fire extinguishers, fire suppressant systems	100	January 1, 1994*
Carbon tetrachloride	Solvents, CFC production, chlorine & pesticides	50	January 1, 1996*
Hydrochlorofluorocarbons (HCFCs)	Home air conditioners, plastic insulation, packaging foam, certain aerosols	15	2040**
Methyl chloroform	Industrial solvents for cleaning metal and electronics	6	January 1, 1996
Methyl bromide	Fumigation of soils, commodities and structures	—	2010***

*Essential-use exemptions apply.
**For developing nations a freeze in HCFC consumption after 2015 and complete phaseout by 2040.
***Use is now to be phased out by 2010 in industrial nations and frozen in developing nations at 1995–1998 levels by 2002.

Source: Council on Environmental Quality. *Twenty-Third Annual Report*. Washington, DC: U.S. Government Printing Office 1992; Hilary French, "Learning from the Ozone Experience." In Lester B. Brown et al., *State of the World 1997*. New York: W.W. Norton, 1997.

the international agreement may impose requirements for which control technology does not currently exist. In 1994 the Protocol's scientific panel reported that substitutes for methyl bromide existed for 90 percent of its uses. Pursuant to that report, the Vienna amendment of 1995 required that methyl bromide use be phased out by 2010. By 1995 over 150 countries had ratified the Montreal Protocol.

As shown in Table 3.1, one complicating aspect of the ozone issue is that the life spans of atmospheric gases such as CFCs and halons are known to be up to one hundred years. Consequently, it is projected that full recovery of the atmospheric ozone will take several thousand years after the emissions of the ozone-depleting gases into the atmosphere are terminated. Although the troposphere abundance of ozone-depleting chemicals peaked in early 1994 and began declining in 1995 (see Figure 3.2), the ozone layer is projected to gradually recover over the coming decades (see Figure 3.3). Only time will tell whether the international system will react fast enough to avert ozone depletion.

Figure 3.2
Tropospheric Abundance of Ozone-Depleting Chemicals from Humans

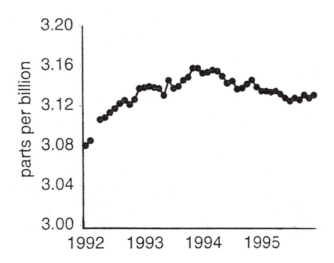

Source: Council on Environmental Quality. *Twenty-Fifth Anniversary Report*. Washington, DC: U.S. Government Printing Office, 1994. Based on data of S. Montzaka et al. "Decline in the Tropospheric Abundance of Halogen form Halocarbons: Implications for Stratospheric Ozone Depletion." *Science* 272 (1996): 1318–22.

The preceding notwithstanding, the Montreal Protocol demonstrates that even when dealing with international law, once scientific consensus is sufficiently well established, public and media concerns can indeed be

Figure 3.3
Adjusted Atmospheric Chlorine Loading

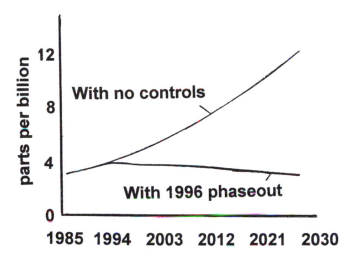

Source: Council on Environmental Quality. *Twenty-Fifth Anniversary Report*. Washington, DC: U.S. Government Printing Office, 1994.

translated into action. Hilary French, a senior researcher at the Worldwatch Institute, puts the Montreal Protocol into a historical perspective:

Future historians may well view the signing of the Montreal Protocol in 1987 as a defining moment—a point at which it became clear that the very definition of international security was undergoing fundamental change. In foreign ministries around the world, the focus has shifted in the decade since Montreal away from such cold war concerns as nuclear arms control treaties and toward the burgeoning domain of environmental diplomacy.[7]

Implications of Global Warming

The possibility of global climatic change, induced by an increase in pollutants in the atmosphere, is potentially the most important international issue facing humanity. Yet, decision makers have a very difficult time dealing with this concern because of the scientific uncertainties, and fighting against global warming may well carry them too far from the immediate social and economic needs of most of their constituents. This section considers the limits and scope of international law in dealing with the global commons. It stresses the scientific uncertainties that emerge when scientists attempt to understand the dynamics of the ecosphere.

The composition of the atmosphere is a major determinant of the Earth's temperature and climate. The process of maintaining the Earth's temperature within fixed ranges occurs, in part, because the sun and the Earth interact to create something that can be likened to a "greenhouse." Thirty percent of the incoming radiation from the sun is scattered or reflected by clouds or by the Earth's surface (global albedo). Twenty percent is absorbed by oxygen, ozone, water vapor and droplets, and dust. The remaining 50 percent of the incoming radiation reaches the ground or ocean where it is absorbed as heat (see Figure 3.4). A very small percentage (less than 1 percent) of the light energy that reaches the surface of the Earth is converted to chemical energy by plants.

The absorbed heat that is finally given off by the Earth does not readily escape into outer space. This heat is absorbed by carbon dioxide and other

Figure 3.4
Relative Strength of Energy Flows and Conversion Rates

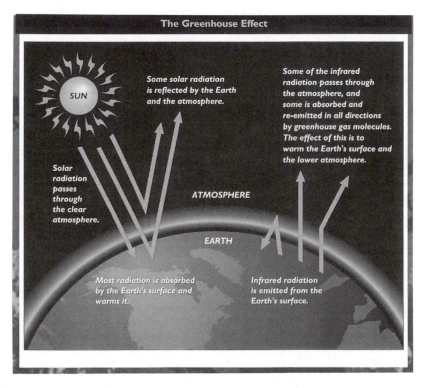

Source: Department of State. *Climate Action Report.* Washington, DC: U.S. Government Printing Office, 1995.

substances in the atmosphere that cause this "greenhouse effect," or heat-trapping effect. That is, the sun, as a high-temperature body, emits short wavelength radiation that is not readily absorbed by these substances. The Earth, as a lower-temperature body, emits long-wave infrared radiation, which is then absorbed by atmospheric substances. A natural "greenhouse effect" maintains the Earth about 33°C (6°F) warmer than it would otherwise be. This same principle occurs on a sunny day in a car with closed windows. Shorter wavelengths of light readily penetrate the windows and warm the seat and dashboard. The longer wavelengths of radiation emitted by the seat and dashboard do not readily pass through the windows. Hence, the inside of the car gets hot (heat-trapping effect).

Recently, humans have been adding more carbon dioxide and other gases to the atmosphere, and have been heating up the environment by their various activities at work, home and play. Another disturbing world-wide trend is the continuing loss of forests to urban and other developments. During photosynthesis, plants incorporate carbon dioxide and give off oxygen. Fewer plants means more heat-trapping carbon dioxide remaining in the atmosphere. The combined effect of greenhouse gases, heat pollution, and deforestation results in an increased warming of the Earth's atmosphere.

Every human activity involves energy. The energy we need to live and work is supplied to us through the food we eat, and is known as dietary energy. We need other fuels as well, such as fossil fuels. Fuel energy is used to reduce physical labor and provide such essentials as light, heat, and transportation. As dietary and fuel energy are burned, much of the energy is dissipated or diluted as waste heat. In highly populated urban centers, this waste heat is enough to cause inner-city temperatures to be 5°F to 20°F warmer than adjacent rural areas (Figure 3.5). On a worldwide scale, however, this waste heat is less significant. Recent studies indicate that in order to raise the global temperature 1°F, humankind would have to increase its energy consumption by a hundredfold.

An increase in carbon dioxide and other gases is expected to have a significant effect on the global temperature. This is because carbon dioxide and other gases (e.g., methane, nitrous oxide, ammonia, sulfur dioxide) slow down the rate of heat loss from the Earth by absorbing infrared radiation. The combustion of fossil fuels is one of the primary causes of increased atmospheric concentrations of carbon dioxide and other gases. Other contributing factors include the burning of forests and changes in the organic levels of soils, both of which are products of deforestation and cultivation.

Figure 3.5
Temperature Changes as a Function of the Density of Trees and Development

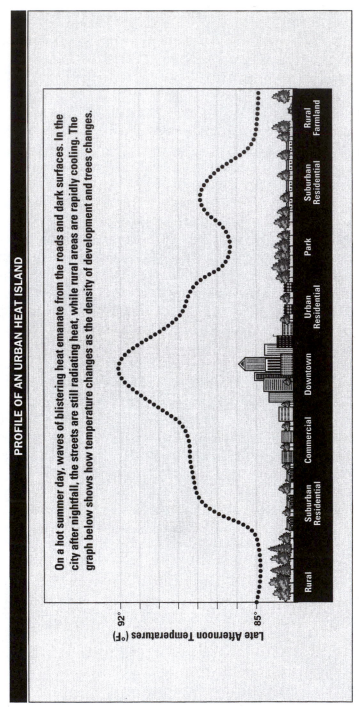

PROFILE OF AN URBAN HEAT ISLAND

On a hot summer day, waves of blistering heat emanate from the roads and dark surfaces. In the city after nightfall, the streets are still radiating heat, while rural areas are rapidly cooling. The graph below shows how temperature changes as the density of development and trees changes.

Late Afternoon Temperatures (°F)

92°

85°

Rural   Suburban Residential   Commercial   Downtown   Urban Residential   Park   Suburban Residential   Rural Farmland

Source: Department of State. *National Action Plan for Global Climate Change.* Washington, DC: U.S. Government Printing Office, 1992.

It has been estimated that atmospheric carbon dioxide levels have increased by 15–25 percent since 1800. As shown in Figure 3.6, carbon dioxide concentration has been increasing linearly since 1958 (a 12-percent increase). Projections of future increases in fossil fuels combustion indicate that the atmospheric levels of carbon dioxide could double over the next hundred years.

Two effects of deforestation are noteworthy. First, changes in the Earth's climate might result from alterations in the reflection of light (global albedo) and heat from the Earth's surface, when light-absorbing forests are removed. (The percentage of the total radiation of a planet that is reflected from its surface is called albedo. As mentioned earlier, the average albedo of the Earth is 30 percent.) The heat balance of the Earth would change, producing consequent changes in wind and rainfall patterns. Second, forests are a great storehouse of carbon; roughly half of the carbon in the Earth's biomass is stored in forests. With large net losses of forests, the concentration of carbon dioxide in the atmosphere could rise by 30 percent, adding to the already rising present trend.

While this global warming is continuing, people's activities are also interfering with the environment by way of emitting pollutant aerosols into

Figure 3.6
Carbon Dioxide Concentration in Air

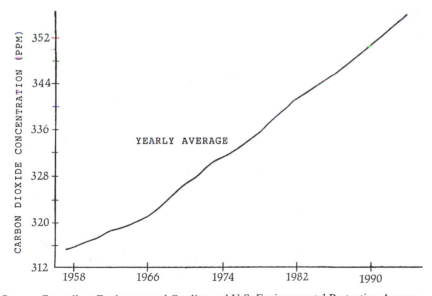

Source: Council on Environmental Quality and U.S. Environmental Protection Agency.

the atmosphere. Scientists warn that this increase of particulate matter in the atmosphere may scatter or reflect back into space about 30 percent more of the sun's radiation than is currently being reflected, resulting in a cooler climate.

If the particulates are making the Earth cooler, but the increased carbon dioxide, other gases, and heat are making it warmer, what direction will the Earth's temperature ultimately take? The cooling of the Earth could speed up the inevitable ice age. On the other hand, a rise of 2°F–3°F in a period of three to four decades could lead to both a thermal expansion of oceans and the melting of the polar ice caps. As a result, sea level would rise and flooding may occur in many coastal and low-lying areas. The sea level might be raised by four hundred feet, and dust bowls and desert temperatures would be reached over most of the world.

According to the data, the average temperature has increased between 0.3°C and 0.6°C over the past one hundred years (Figure 3.7). A number of recent studies suggest the possibility of global warming between 1.5°C and 4.4°C within fifty years, if no actions are taken to limit greenhouse gas emissions. However, environmental effects associated with climatic change would not be wholly adverse. Due to temperature change, productivity in high latitudes would increase (because of a longer growing season). Nevertheless, in the interior mid-latitude regions, such as the midwestern United States, the former Soviet Union, and China, potentially drier conditions coupled with higher temperatures could reduce productivity.

Figure 3.7
Global Mean Annual Surface Temperature

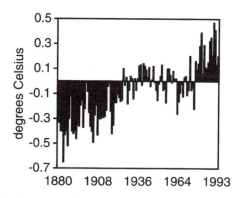

Source: Council on Environmental Quality. *Twenty-Fourth Annual Report*. Washington, DC: U.S. Government Printing Office, 1993.

New studies indicate that global warming may exacerbate the spread of tropical diseases into temperate regions in the next few decades. It is also projected that due to global warming, the tropical carriers of certain diseases, such as mosquitoes and flies, will increase their abundance and distribution. Thereby, it is expected that there will be an increase in the death rate of the world population due to malaria, cholera, sleeping sickness, yellow fever, bronchitis, asthma, and many other ailments.

To assess the current state of knowledge regarding human-induced changes in the Earth's climate and possible consequences, the United Nations convened an international group of scientists, economists, and decision theorists. In its latest assessment, the group, called the U.N. Intergovernmental Panel on Climate Change (IPCC), concluded in its landmark 1995 report that "the incidence of floods, droughts, fires and heat outbreaks is expected to increase in some regions" as temperature increases.[8] However, developing nations, which have contributed little to global warming, will suffer the most from climate change (see Figure 3.8). This point is elaborated on by Christopher Flavin of the Worldwatch Institute:

Developing countries are particularly vulnerable to climate extremes since many have high population densities and cannot afford to protect farmland or homes, or even evacuate threatened areas expeditiously. In countries such as Bangladesh, where millions have no choice but to live and farm in areas vulnerable to flooding, the results could be devastating. Moreover, people in developing countries generally do not have insurance policies that would compensate them once a disaster is over.[9]

Many islands are particularly subject to flooding from a rise in sea level. The former president of the island republic of Kiribati, Ieremia Tabai, has

Figure 3.8
Global Carbon Dioxide Emissions by Region, 1992

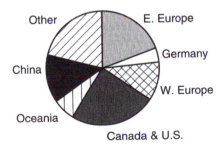

Source: Council on Environmental Quality. *Twenty-Fourth Annual Report.* Washington, DC: U.S. Government Printing Office, 1993.

rightly expressed concern about global warming. He said: "If the green-house effect raises sea levels by one meter it will virtually do away with Kiribati. . . . In 50 or 60 years, my country will not be here."[10]

Scientists have long debated how the increasing levels of carbon dioxide in the atmosphere will affect the temperature of the Earth. Too much is still unresolved, including the interaction of the effects of past air emissions, the risk of future air emissions, the type of pollution response curve between sources and receptors of present air emissions, and its socioeconomic factors. The latter includes the relative costs and benefits of various alternative courses of action (Figure 3.9). One should note that, even if control measures are fully implemented, global climatic change will continue for nearly another century. Because of the long atmosphere life spans of many greenhouse gases (up to one hundred years), a return to near-natural levels of these gases will take centuries, if they are at all recoverable.

As mentioned earlier, the atmosphere's carbon dioxide concentration increased by 12 percent since 1958. This was largely caused by deforestation and the burning of vegetation and fossil fuels. If we continue to add airborne carbon dioxide at our current rate, the planet's average temperature will rise, but it is uncertain by how much (perhaps between 1.5°C and 4.4°C). This increase is predicted to occur by the middle of the twenty-first century. Moreover, different computer models disagree on how each region of the world will be affected. Major sources of uncertainties are the roles of the atmosphere, oceans, photosynthetic organisms, and soil.

The models estimate that about 55–65 percent of the carbon dioxide emissions generated by fossil fuels are presently retained in the atmosphere. It is reasonably well-known that only marine environments are currently the significant absorbers of airborne carbon dioxide, whereas terrestrial ecosystems are approximately neutral components (i.e., absorbing as much as they are generating). As a marine ecosystem warms, it will very likely absorb less airborne carbon dioxide, unless photosynthesis increases proportionally. If the removal of trees and other photosynthetic organisms continues, photosynthetic organisms would also represent a reduced resource for absorbing airborne carbon dioxide. These two unknowns can have an impact on the ability of scientists to predict rates of climatic modification.

The difficulty in understanding ocean physics and cryospheric relations is also related to the scientific uncertainty that still surrounds the greenhouse question. Even though it is speculative that alpine glaciers and polar ice packs would melt, the sea level would rise simply because water ex-

Figure 3.9
Actions for Facilitating Adaptation to Rise in Sea Level

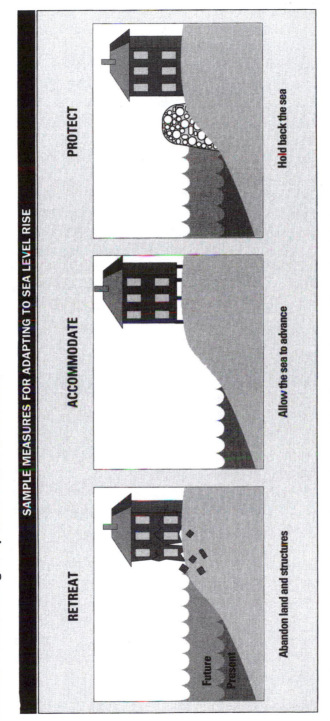

Source: Department of State. *National Action Plan for Global Climate Change.* Washington, DC: U.S. Government Printing Office, 1992.

pands when it absorbs heat. If the Arctic should melt substantially, it is expected that most of the world's low-lying areas would be flooded.

To make food, a plant must obtain carbon dioxide. It is well established that when carbon dioxide levels are raised in chambers, such as greenhouses, most vegetation responds with increased growth. Nevertheless, the potential effect of high carbon dioxide concentrations on agriculture and other plant resources remains largely uncertain. Scientists cannot predict what effects rising carbon dioxide concentrations will have on vegetation in natural ecosystems. One thing is certain, however: not all plants will respond in a similar manner. Consequently, species competition would be influenced, resulting in changes in the structure and functions of ecosystems.

There are a number of reasons why international agreements on greenhouse gases are so difficult to achieve. Controlling these gases requires a massive change in economic priorities, including technology-forcing features and alternative sources of energy. In contrast, the prevention of ozone depletion is relatively easier because it requires less economic and technological change. Ozone-depleting chemicals were produced mainly by a relatively small number of rich nations. Moreover, all nations will not be affected to the same degree if global warming occurs. Indeed, "Some countries may even anticipate gains from warming and changes in rainfall, or they may conclude that the drawbacks of warming would be less burdensome than the costs of prevention."[11]

Some nations, like the United States, have actively shown their commitment to comprehending the events that control the "greenhouse effect." They do this by funding research that aims to minimize the scientific uncertainties that presently exist concerning the magnitude of projected temperature changes for different global regions. However, meaningful efforts to minimize global climatic modifications require binding international laws. Although virtually all nations have expressed interest in international negotiations concerning minimizing global warming, as with so many other major environmental concerns facing international policymakers, the timing and fortitude of their responses are shortsighted.

This point is illustrated by observing the results of the biggest and most important environmental conference—the U.N. Conference on Environment and Development (UNCED). This conference, also known as the "Earth Summit," brought together six thousand delegates from over 170 nations for discussions and negotiations on such topics as climate change, biodiversity, deforestation, and economic relations between developed and less-developed nations.

Held June 3–4, 1992, in Rio de Janeiro, Brazil, the conference produced the following key principles:

1. People are entitled to a healthy and productive life in harmony with nature.
2. Nations have sovereign rights to exploit their resources in accordance with their environment and environmental policies, providing they do not harm the environment elsewhere.
3. Environmental protection must be an integral part of the development process.
4. Unsustainable patterns of development must be reduced, and appropriate demographic policies must be promoted.
5. Human beings have a right to information and the opportunity to participate in political processes.
6. Polluters should pay for their environmental damages through internalization of costs, such as user fees, emission charges, and taxes.

Unfortunately, these lofty principles are not legally binding. Thus, the conference was not as fruitful as it might have been. Many developing nations stonewalled the rush into a legally binding treaty that might restrict the use of their sovereign and revenue-producing forest resources.

On a positive note, a consensus did emerge to negotiate a nonbinding, authoritative statement of principles for the management, conservation, and sustainable development of all types of forests. At the Earth Summit, the United States joined other countries in signing the Framework Convention on Climate Change, an international agreement to address the danger of global climate change. The principal thrust of the Convention was clearly expressed in its second article:

The ultimate objective of this Convention . . . is to achieve . . . stabilization of greenhouse gases concentration in the atmosphere at a level that would prevent dangerous anthropocentric interference with the climate system. Such a level should be achieved within a time frame sufficient to allow ecosystems to adapt naturally to climate change, to ensure that food production is not threatened, and to enable economic development to proceed in a sustainable manner.[12]

The most visible opponent to a legally binding treaty was President George Bush. He resisted calls for action on the global warming treaty, the biodiversity treaty, the forest protection convention, and the promise to donate 0.7 percent of the U.S. gross domestic product to developing nations for environmental protection. President Bush claimed that too much scientific uncertainty exists concerning global warming and that such a treaty would restrict burning of fossil fuels and stifle economic growth. The

United States, which is endowed with a large coal reserve, is concerned about the effects that a clean fuel mandate would have on its energy independence.

President Bush's arguments rested in part on the conclusion from the 1989 Economic Report to the President, which contended that "there is no justification for imposing major costs on the economy in order to slow the growth of greenhouse gases emissions." The report indicated that the costs of achieving just a 20-percent decline in discharges of carbon dioxide in the United States would range from $800 billion to $3.6 trillion.

The Bush administration also claimed that it was already taking action to protect the environment by phasing out ozone-depleting chemicals as part of the Montreal Protocol, which would hold its net contribution of greenhouse gases to 1990 levels by the end of the century.

Against the advice of many scientists and advisors, President Bush convinced several influential nations to support a weakened version of the Rio de Janeiro agreement. The U.S.-sponsored treaty called for countries to assess their greenhouse gas emissions, and required that developed nations submit national plans aimed at reducing their discharges of carbon dioxide and other greenhouse gases to 1990 levels by the year 2000.

However, to alleviate the global warming issue, the treaty requires that industrialized nations provide assistance to developing nations, both in collecting data on net greenhouse gas emissions and in limiting the growth rate of emissions. The treaty further provides for financial aid to developing nations for certain incremental costs of projects that produce global environmental benefits.

The treaty does not require nations to stabilize or reduce greenhouse gas emissions. The treaty did not adequately consider what nations must do beyond the year 2000. Thus, the treaty accommodates variations in the political and economic circumstances of countries, specifically avoiding uniform emission targets and timetables. In short, no binding international agreement came out of the so-called Earth Summit.

On Earth Day 1993, President Bill Clinton stated:

We must take the lead in addressing the challenge of global warming that could make our planet and its climate less hospitable and more hostile to human life. Today, I reaffirm my personal and announce our nation's commitment to reducing our emissions of greenhouse gases to their 1990 levels by the year 2000. I am instructing my administration to produce a cost-effective plan . . . that can continue the trend of reduced emissions. This must be a clarion call, not for more bureaucracy or regulation or unnecessary costs, but instead for American ingenuity and creativity to produce the best and most cost-efficient technology.[13]

Later, in the spring of 1994, President Clinton pledged that the United States would curtail greenhouse gases to their 1990 levels by the year 2000, but spoke of no commitment with regard to limiting discharges after the year 2000. Moreover, reduced congressional funding has limited President Clinton's effectiveness. According to the Council on Environmental Quality, "It appears likely that the U.S. will fall short of the original goal of returning net greenhouse gas emissions to 1990 levels by the year 2000."[14]

In 1995 environmental leaders from about 120 nations met at Berlin for the first Conference of the Parties to the Framework Convention on Climate Change. The second conference convened in Geneva, Switzerland, in 1996, and the third convened in Kyoto, Japan, in 1997.

Surprisingly, the conference in Kyoto reached a framework for handling climate change in coming decades. Delegates from over 150 nations decided to move ahead with international controls in the absence of conclusive proof of climate change. Under the terms of the Kyoto Protocol, thirty-eight industrial nations are required to cap their emissions of climate-altering gases below 1990 levels for the period 2007 through 2012. Japan must reduce its emissions by 6 percent, the United States by 7 percent, and the 15-nation European union by 8 percent. Some countries, because of their economic circumstances, would face smaller cuts, and some would not face any at the present. As a group, the nations would cap emissions of carbon dioxide, methane, nitrous oxide, and three halocarbons by slightly more than 5 percent below 1990 levels. Issues concerning trade emission credits and other details were negotiated in the next round of climate discussions in Buenos Aires, Argentina, in November 1998. Over 150 nations set operational rules for reducing emissions of greenhouse gases.

These conventions are merely preliminary arrangements for a formal treaty on global warming in the hope that legally binding treaties similar to the Montreal Protocol will emerge from these international conferences on global warming. But it is worth remembering that international laws are inherently dysfunctional and can be challenged on several grounds. For example, treaties bind only those countries that voluntarily agree to comply with them, and no true executive authority exists to enforce them. Many developing countries are already denouncing the Kyoto Protocol, saying that they will not act to reduce climate-altering gases until developed countries comply with the accord. Moreover, the developing countries delayed a compromise that would allow polluting countries to purchase or trade emission credits from countries whose emissions are below the acceptable levels. On the other hand, many industrialized countries, particularly the United States, are insisting that the treaty should impose requirements on newly industrialized nations and require developing countries to participate in a global plan to reduce climate-altering gases. During the next few years, people worldwide will be subjected to lobbying by powerful coalitions (environmentalists versus fossil-fuel providers and industries) that support and oppose the proposed treaty.

The preceding notwithstanding, several positive developments have resulted from these international conferences. The first is that they brought attention to the endangered global commons. Secondly, the conferences brought together international decision makers. Finally, the conferences galvanized the world's emerging environmental movement.

The importance of the world's environmental movement should not be overlooked. Most environmental agreements are not yet subject to international adjudication, and other mechanisms may be used to enforce them. Some of these agreements have been enforced by means of trade measures, pressures from nongovernmental organizations, and debt-for-nature swaps. International commitment to protecting the global commons, an effort that has attracted the attention of both public and private decision makers, has demonstrated the value of widespread cooperation in the affairs of government.

NOTES

1. Council on Environmental Quality, *Environmental Quality: The Sixteenth Annual Report* (Washington, DC: Superintendent of Documents, 1985), 218.

2. Rachel Carson, *Silent Spring* (Boston: Houghton Mifflin, 1962), 244.

3. Robert C. Mitchell, "From Conservation to Environmental Movement," in Michael J. Lacey, ed., *Government and Environmental Politics* (Baltimore: The Johns Hopkins University Press, 1991), 84.

4. Barry Commoner, *The Closing Circle* (New York: Alfred A. Knopf, 1971), 38.

5. Garrett Hardin, "Tragedy of the Commons," *Science* 162 (1968): 1244.

6. Council on Environmental Quality, note 1, 224.

7. Hilary F. French, "Learning from the Ozone Experience," in Lester B. Brown et al., eds., *State of the World 1997* (New York: W.W. Norton, 1997), 168.

8. Intergovernmental Panel on Climate Change, *Assessment of Knowledge Relevant to the Interpretation of Article 2 of the UN Framework Convention on Climate Change* (New York: United Nations, 1995).

9. Christopher Flavin, "Facing Up to the Risks of Climate Change," in Brown et al., eds., *State of the World 1996*, p. 28.

10. Cited in Colin D. Woodroffe, "Preliminary Assessment of the Vulnerability of Kiribati to Accelerated Sea Level Rise," in Joan O'Callahan, ed., *Global Climate Change and the Rising Challenge of the Sea* (Silver Springs, MD: NOAA, 1994).

11. Oscar Schachter, "The Emergence of International Environmental Law," *Journal of International Affairs* 44 (1991): 473.

12. Cited in U.S. Department of State, *National Action Plan for Global Climate Change* (Washington, DC: U.S. Government Printing Office, 1992), 10.

13. William J. Clinton and Albert Gore, Jr., *The Climate Change Action Plan* (Washington, DC: U.S. Government Printing Office, 1993), i.

14. Council on Environmental Quality, *The Twenty-Fifth Anniversary Report* (Washington, DC: U.S. Government Printing Office, 1995), 218.

4

The Environmental Concern for Overpopulation

Presently, the human population is increasing exponentially at the rate of approximately 1.5 percent annually. If this growth rate were to continue, one can imagine that the sheer mass of all living humans, in a few thousand years, would be greater than the mass of the Earth. In order for the human biomass to mushroom to this level, it would need to devour the Earth itself. Planet Earth is essentially a closed system with respect to matter. There is no transfer of matter between the Earth and its surroundings. Because the number of atoms on Earth is finite, a species grows in biomass at the expense of its surrounding environment by obtaining atoms from the Earth. Consequently, the human biomass can never weigh more than the Earth; unless atoms are obtained from other planets, exponential growth cannot occur on Earth forever.

This population explosion, in many respects, is one of the rudimentary causes of environmental problems. Holding all other variables constant, humans will eventually affect the environment; larger populations will consume enormous quantities of resources and will subsequently generate more pollution. With a zero population growth, society could concentrate on improving environmentally benign technology while raising the quality of goods and services. On the other hand, with an expanding population, society must make use of its resources by providing new goods and services for the growing population. For instance, instead of building factories, resources could be diverted to make the present ones more efficient.

The emergence of the third wave of environmentalism involves humanity's awareness of its biomass, and the antipopulation policies that are sparked by this concern. Because the population crisis may increase demand for environmental laws, which can sometimes infringe on reproductive rights, population policies are generally topics that politicians are not willing to discuss in public. This is not to say that the population issue has not come up indirectly in other ways—for instance, in funding for famine relief and aid for family counseling. The U.S. Congress has never debated the need for zero population growth. To further illustrate, the overpopulation problem was barely mentioned in the "Earth Summit" document at Rio de Janeiro due to the opposition from religious groups.

Unlike "biosphere conservation" (Chapter 2) and "pollution prevention" (Chapter 3), population control is not a "mother and apple pie" issue. In fact, the vast majority of environmentalists advocate wilderness preservation and antipollution policies rather than population control.

A number of population alarmists have proposed their brand of population policies to deal with overpopulation and resource scarcity in society. During the early stage of the population movement, two brothers, Paul and William Paddock, the former an agronomist and the latter a foreign service officer of the State Department, in 1967 proposed a draconian measure known as the triage method.[1] Under this scenario, nations are divided into three groups: (1) nations capable of dealing with their population problems, (2) nations that could deal with population problems but require foreign assistance, and (3) nations that are truly in an extreme predicament, where even foreign aid cannot help them to achieve self-sufficiency. Under the triage policy, when environmental resources are severely limited, foreign aid is provided to solve population problems for nations that can be saved, such as those in group two. Unfortunately, nations in group three are left without foreign assistance. The triage model is a harsh and inhumane method in dealing with the overpopulation problem. But, as discussed later, this extreme public policy may not be necessary.

During the early stages of the environmental movement, one event that heightened the concern of overpopulation into prominence and created a storm of controversy was the 1968 publication of *The Population Bomb* by Paul Ehrlich. This book was written for the general public and was widely read by many individuals. It immediately became an environmental classic, selling hundreds of thousands of copies in just a few months. In his book, Ehrlich argues that most environmental problems are the result of overpopulation. Ehrlich starkly warned:

In order to just keep the standard of living at the present inadequate level, the food available for the people must be doubled. Every structure and road must be duplicated. The amount of power must be doubled. The capacity of the transport system must be doubled. The number of trained doctors, nurses, teachers, and administrators must be doubled. . . . This would be a fantastically difficult job in the United States—a rich country with a fine agricultural system, immense industries, and access to abundant resources. Think of what it means to a country with none of these.[2]

One year after the publication of *The Population Bomb*, U Thant, the Secretary-General of the United Nations, further publicized the issue of overpopulation and gave it an official recognition by apocalyptically saying:

I do not wish to seem overdramatic, but I can only conclude from the information that is available to me as Secretary-General, that the Members of the United Nations have perhaps ten years left in which to subordinate their ancient quarrels and launch a global partnership to curb the arms race, to improve the human environment, to defuse the population explosion, and to supply the required momentum to development efforts. If such a global partnership is not forged within the next decade, then I very much fear that the problems I have mentioned will have reached such staggering proportions that they will be beyond our capacity to control.[3]

The ideology of Paul Ehrlich and other like-minded individuals can be traced back to Thomas Malthus, an eighteenth-century clergyman, political theorist, and economist. Malthus observed that population increase among the poor was exponential (1, 2, 4, 16, . . .), whereas food-growth increase, at best, was an arithmetical progression (1, 2, 3, 4, . . .). Consequently, the living standard can never rise much above the subsistence levels because of the constant population pressures placed on the food supply. As you can see in Figure 4.1, because population growth surpasses food production, the results that Malthus predicted were inevitable famines, plagues, and wars.

In this chapter, we examine the third and the most controversial stage of the environmental movement, namely, the concern about overpopulation. We study the population patterns and the policies of developed and developing nations. In order to begin our study, we must first consider some general properties of population ecology. An understanding of relevant concepts of population ecology will help society appreciate the ways in which the overpopulation problem might be solved, and the ways in which society can be nudged into a sustainable population growth rate. Let us briefly see what lessons can be learned from population ecology.

Figure 4.1
Exponential or Geometric Increase versus Arithmetic Increase

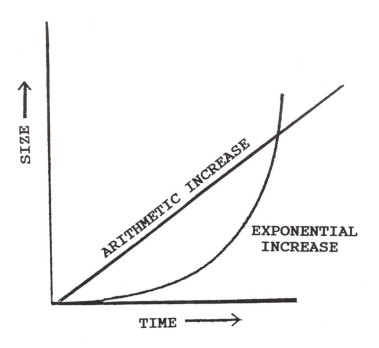

LESSONS FROM POPULATION ECOLOGY

Anyone who has taken a course in ecology knows that the number of individuals in a species inhabiting a specific area rarely indicates how many there would be, should the environment be favorable. This is because a population has an exponentially reproductive potential. For example, female house mice are dependent upon their parents for three to four weeks after birth but become sexually mature at seven weeks of age. If given the opportunity, each female might produce eleven litters in ten months, totaling to one hundred or more offspring. Therefore, within a four-year period, a pair of male and female mice and their offspring could produce a total of over ten million mice. A few hundred years later, the mass of the mouse population would be larger than that of the Earth (see Figure 4.2).

According to ecologists, when resources are abundant and there are no predators or other factors controlling a population, the specific growth rate (the population growth per individual) becomes constant and maximized for the existing microclimatic conditions. Ecologists use the term "biotic potential" to designate the maximum reproductive power. Due to this biotic

Figure 4.2
Exponential Population Growth

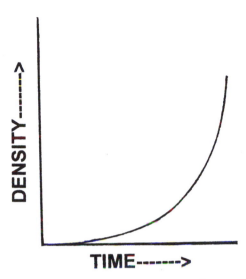

Source: Adapted from Richard Brewer. *The Science of Ecology.* Ft. Worth, TX: Harcourt
Brace, 1994.

potential, the population of a species could double at an accelerating rate, as
shown in Figure 4.2.

Human beings are not an exception. There is also great potential for rapid
population growth in human beings. As demonstrated in Figure 4.3, the human population increased at a relatively slow rate until almost 1650, at
which time approximately half a billion people inhabited the Earth. The
population then doubled to the first billion within the next two centuries, the
second billion was added within the next eighty years; the third billion
within the next thirty years; the fourth billion within the next fifteen years;
the fifth billion within the next twelve years; and the sixth billion will be
added within the next eight years. If this growth rate continues, the time period between each new billion will get shorter and shorter. Thus, these increasingly shorter periods of time indicate that humans are also capable of
exponential growth.

As shown in Table 4.1, the doubling time differs from one nation to another. However, in most cases, the reproductive potential of these nations
would eventually cause them to double in size; only the time span necessary
to achieve this doubling of population differs. The lesson here is simple and
of vital importance: As long as the birth rate exceeds the death rate, regard-

Figure 4.3
Growth Curve for Human Population: Past (solid line) and Projected (broken line)

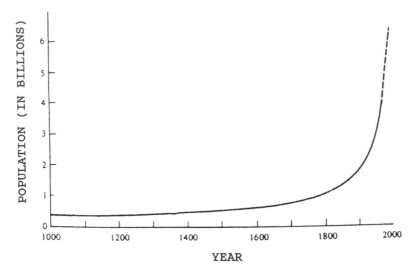

Source: Population Reference Bureau. *World Population Data Sheet*. Washington, DC: Population Reference Bureau, 1995.

less of how small a difference there is, the population will double. This growth rate, when plotted into a graph, produces an exponential curve (as in Figure 4.2). Moreover, it is unlikely that the population of the most-developed nations will ever double under the current stage of demographic dynamics. As explained later in this chapter, both birth rates and death rates reflect social factors and show a strong correlation to the age factor. The global population may reach the eleven to twelve billion range by the end of the twenty-first century.

To reiterate, at the present growth rate of 1.5 percent, one can theoretically determine that the sheer mass of all human inhabitants will be equal to the mass of Earth in a few thousand years. If the growth rate stays the same, in a few million years, the human biomass would weigh as much as the visible universe, and would be "expanding outward at the speed of light." This biotic potential is purely a biological goal, which is not different from that of any other plant or animal species of Earth. Excess reproduction is the means by which species compensate for the great number of offspring that fail to survive due to predation, parasitism, starvation, and so forth.

Table 4.1
Comparison of Basic Demographic Data in Poor, Moderate, and Rich Nations in Terms of Economic Development

	Population 1996 (Millions)	Average Life Expectancy (Years)	Annual Rate of Growth (in %)	Doubling Time (Years)
Afghanistan	22.7	45.5	2.4	29
Bangladesh	128.1	55.5	2.3	30
Cambodia	10.8	49.5	2.8	25
Chad	6.9	41.0	2.2	32
Ethopia	57.0	50.0	3.1	23
Haiti	6.7	45.0	2.0	35
Kenya	28.8	52.5	3.0	23
Mozambique	18.1	49.0	2.9	24
Rwanda	8.6	39.0	2.7	26
Somalia	9.6	55.5	3.2	22
Brazil	161.0	62.0	1.2	58
India	937.0	59.0	1.8	39
Indonesia	204.0	61.0	1.6	44
Mexico	94.0	73.5	2.2	32
Peru	24.0	66.0	1.8	39
Thailand	60.0	68.5	1.2	58
France	58.0	78.0	0.4	175
Germany	81.0	76.5	0.0	
Japan	126.0	79.5	0.3	233
Sweden	9.0	78.5	0.2	350
Switzerland	7.0	78.5	0.3	233
United States	264.0	76.5	0.7	100

Generally, degenerative diseases of the aged, such as heart problems and cancer, are far more important as the causes of death among developed nations while communicable diseases are more significant in developing nations.

Source: Robert Famighetti, ed. *The World Almanac*. Mahwah, NJ: Funk & Wagnalls, 1996.

Humans are the only fast-growing species that can substantially alter their environment. Apparently, exponential growth cannot occur indefinitely. In nature, large population growth of all wildlife is prevented by five factors: resource availability, consumers (for example, predators, parasites), competition, self-regulation, and catastrophic events. Collectively, these factors opposing unlimited growth are known as environmental resistance. In modern societies, the factors that tend to limit population size include pollutants that occur principally in developed nations and/or with food production, which is mostly associated with developing nations.

The carrying-capacity principle states that as the population approaches the optimum level of sustainable size, or carrying capacity, environmental resistance becomes greater and greater. The carrying capacity is determined by available resources and other limiting factors in a given area. The difference between biotic potential and the actual population growth is a measure of environmental resistance (see Figure 4.4).

Ecologists have distinguished two general types of growth curves, the S-shaped and J-shaped. The S-shaped growth curve is produced when envi-

Figure 4.4
The Environment's Ability to Support an Expanding Population of Any Organism over an Extended Period of Time

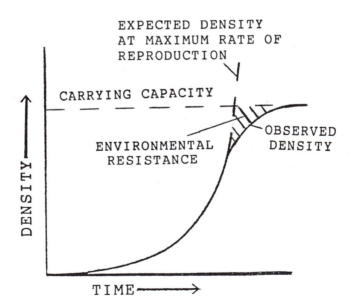

Source: Adapted from Richard Brewer. *The Science of Ecology*. Ft. Worth, TX: Harcourt Brace, 1994.

ronmental resistance becomes increasingly effective as the density of the population rises (Figure 4.5). The environment resistance becomes density dependent, or the intensity of the resistance is partly determined by the density of the population. The rate of increase slows down as the population grows older and approaches the optimum population of individuals that the habitat and/or nutritional resources can support. When the reproduction rate equals the death rate, the overall number of individuals remains constant. This is referred to as the stationary phase. In some species, such as the fruit fly and paramecium, the population levels off at the carrying capacity. In others, such as bacteria and yeast, the exhaustion of nutrients in the environment and/or the accumulation of toxic wastes may cause a deceleration of the population growth.

The J-shaped curve is a characteristic of the population growth of small insects with short life cycles and of annual plant populations. In such a curve, the population increases in density at a rapidly accelerating rate until the environmental resistance suddenly comes into play. At this point, the

Figure 4.5
The Typical S-shaped Population Growth Curve

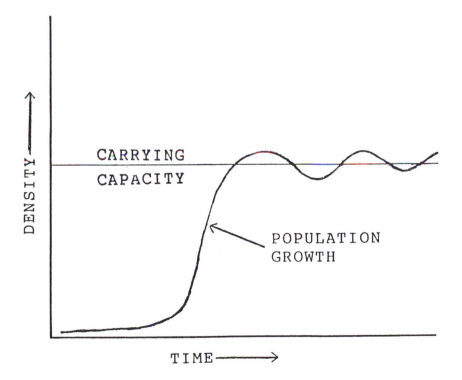

population density falls dramatically (see Figure 4.6). In this growth pattern, the population grows exponentially until it strikes the limit set by the habitat and/or nutritional resources. Thereafter, the population suffers a reduction that is independent of its density. During deceleration, the population declines abruptly due to exhaustion of available resources.

Are there any historical examples of the J-shaped curve for people? We know, from the whole course of human history, that food is probably the most important controlling factor and is forever causing the population to decline. Examples are the legions that died in 1,828 famines in China from 108 B.C. to A.D. 1811; 201 in the British Isles; 2 in the former Soviet Union (1 from 1921 to 1923 and 1 from 1932 to 1934); and today famine in Africa and the Near East, just to name a few.

The last major famine to strike Europe occurred in Ireland. The fundamental cause of this catastrophe was population pressure on the available land and also bureaucratic blunders. The population of Ireland grew exponentially from about 800,000 in 1500 to 8.5 million in 1846. During

Figure 4.6
The Typical J-shaped Population Growth Curve

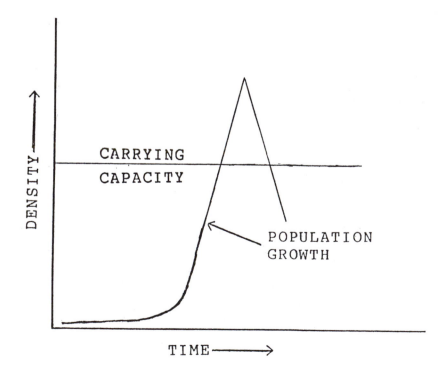

1845–1849, the potato blight, a destructive fungus disease, devastated Ireland's potato harvest. British and other European governments did not act swiftly and effectively to save the famine victims. About 1–2 million people died directly or indirectly from undernourishment and malnutrition as a result. Another 1–2 million people emigrated, mostly to the United States and Canada.

Every species displays a tendency to increase in number. Some species follow the S-shaped curve, others the J-shaped curve, and the rest are a mixture of the two. Even though the environment has neither the habitat resources nor the nutritional resources for all the possible offspring of any species, the inherent readiness to overreproduce is an innate characteristic of every species. The only variable involved is the time it takes to reach a large population size. Because there are many causes of death in nature—including predators, parasites, floods, starvation, and diseases—excess reproduction is valuable to a species when many offspring fail to survive to maturity.

Is our growth attitude largely genetic? Some researchers have observed that "family size has a tendency to run in families."[4] Perhaps the biblical mandate to "be fruitful, and multiply, and replenish the earth" (Genesis 2:28) originated from our biological requirement to survive as a species. If so, environmental laws may regulate and influence the reproductive instinct, but they can never repeal it. The legislature cannot rationally enact laws commanding that the reproductive instinct should not exist. Consequently, society must find ways of balancing humanity's fundamental drive to reproduce against environmental deterioration.

Similar to other species, humans also have encountered environmental resistance. Unlike other species, they have had the ingenuity to avoid or postpone the effects of environmental resistance. As a result, the human population has been spurred on by three "revolutions," each conquering some form of environmental resistance.

The cultural revolution may have been the first shift that allowed the human population to enter a phase of exponential growth. This included the discovery of fire and the development of tools, weapons, dwellings, and protective clothing. With these discoveries, primitive people were able to modify and exploit their environment more effectively. The discovery of weapons and fire opened up a new food range. Cooking, for example, kills food pathogens, removes volatile plant toxins, and softens plants by making food more digestible. Also, with the discovery of protective clothing and shelter, humans were capable of penetrating and establishing themselves on all the continents. Ten thousand years ago, the Earth probably supported about five million humans.

Plant cultivation and animal husbandry may have allowed the second surge of growth. About ten to twelve thousand years ago, people domesticated plants and animals. For the first time, they learned how to breed them. They also learned to cultivate, till, fertilize, and irrigate the soil. According to some estimates, by domesticating plants, humans enormously reduced the land required for sustaining each person by a factor of the order of at least 500. As a result of the increase in dependable sources of food that became available, the human population began to grow rapidly, climbing to an estimated 130 million people by the time of Christ.

The third population growth surge began in Europe between the sixteenth and seventeenth centuries. The factors contributing to this and the continued growth in the twentieth century are not fully known for all the regions of the world. However, in many regions, it is highly probable that three major factors were involved: industrial, agricultural, and medical. These factors, underpinned by cultural factors, may have spurred the third surge in population growth.

The Renaissance in Europe during the fourteenth and fifteenth centuries, which renewed interest in science and medicine, along with the "age of discovery" in the sixteenth and seventeenth centuries, which introduced new food plants to Europe such as corn (maize) and the potato, ultimately led to the establishment of industry, improved medicine and sanitation methods, as well as other means of disease control. Farming techniques were improved, and crop failures were less frequent. The discovery of antibiotics and the development of many new vaccines in the twentieth century further reduced major limiting factors of the human population growth. This population "explosion" continues today, and in many respects, is one of the reasons for the current environmental crisis. As described in Chapter 2, many analysts believe that this scientific-technological revolution was based on the Judeo-Christian proposition that nature exists to serve humankind.

In the future, humans are likely to conquer the present-day environmental resistance, and experience a fourth population growth surge. The only questions are When? Where? How? However, the most important issue facing humanity now is whether population will be able to level off at the present carrying capacity in order to avoid human suffering. With these key concerns in mind, we now turn our attention to a study of world population growth.

WORLD POPULATION GROWTH

Figures 4.3 and 4.7 show an approximation of the history of human population size. Eight thousand years ago, the world was inhabited by about

5–10 million humans. This number had slowly and steadily increased until two thousand years ago, when the population stood between 100 and 160 million. By the late eighteenth century, the number had grown slowly and unsteadily, increasing to just less than 1 billion. With the arrival of the industrial revolution, the trend changed to a more stable increase. By the end of the twentieth century, the Earth's population will increase to 6.2 billion. The present population represents an apparent exponential stage. From our earlier discussion of growth curves, we can speculate that it is either a J-shaped curve or the first part of an S-shaped curve. It is impossible to gauge whether the worldwide growth curve is a J- or an S-shaped curve, because the population growth differs from one nation to another (Table 4.1). This variability is illustrated by the growth pattern of industrialized regions in North America and Europe (Figure 4.7). As a rule, these developed regions increased their population sizes principally due to the decline of the death rate. This decline is associated with the technological development that has in turn sparked an increase in food production, and a reduction of the effects of the physical environment. It also made parasites less abundant and im-

Figure 4.7
A Comparison of World Population by Region

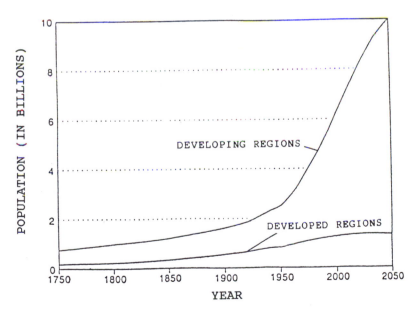

Sources: United Nations Population Division. *Long-Range World Population Projections*. New York: United Nations, 1992; World Resource Institute. *World Resources 1994–1995*. New York: Oxford University Press, 1995.

proved medical care. Although these nations have exhibited some self-regulatory mechanisms like warfare, there has been no significant effect on their population trends.

Developed nations have experienced a greater reduction in growth rates compared to the developing nations. This reduction can be attributed to a decline in the fertility rate because the people in these regions are opting mostly for smaller families. In the 1940s, demographers observed that a decline in a developed nation's birth rate eventually followed a reduction in its death rate. This reduction in birth rate is known as the demographic transition.

In the demographic transition, a nation's growth rate goes through three phases. In the first phase, the growth rate is low due to a high death rate (especially among infants), which effects the high birth rate (Figure 4.8). In the second phase, as a nation begins to develop, the growth rate is high because of a lower death rate. The result is that there is a surge in the population. In the third phase, the growth rate is low again because the birth rate declines.

Figure 4.8
Demographic Transition

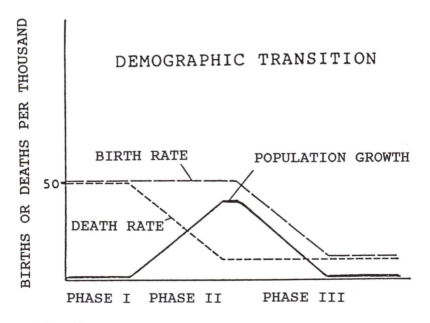

Sources: Adapted from Wentworth B. Clapham. *Human Ecosystems*. New York: Macmillan, 1981; Bernard J. Nebel. *Environmental Science: The Way the World Works*. Englewood Cliffs, NJ: Prentice-Hall, 1987.

The population enters a state of equilibrium, which means that the population may oscillate somewhere above or below the carrying capacity.

The reasons for the correlation between economic development and low birth rates are poorly understood, and are not mutually exclusive. Some scholars attribute this trend to industrialization and urbanization. As more people settle in cities looking for jobs, children are no longer as beneficial as they were in rural settings. From an economic perspective, children represent a serious drain on their parents' pocketbooks. Another hypothesis states that living in the city makes people more cognizant of the concerns of overcrowding, and thus the birth rate decreases.

In addition, some reports indicate that there is some correlation between family size and socioeconomic status.[5] They suggest that as the developed nations become more affluent, infant mortality declines, and educational levels increase, then the birth rate would decrease. To illustrate this point further, note that in Table 4.1 the nations with higher life expectancy were also more affluent and tend to have lower annual rates of population growth than nations with lower life expectancy. Another reason given for the decline in birth rates is that urban living combined with industrialization provides more career alternatives for women. The availability of reliable birth control devices, especially with the advent of "the pill" in the United States in the 1960s, also made family planning more realizable.

Regardless of the rationale for the declining growth rate, the fact remains that populations have more or less stabilized in most developed regions. Some examples are countries such as Australia, Belgium, Denmark, Finland, France, Italy, Luxembourg, Norway, Sweden, and the United Kingdom—all of which are virtually at zero population growth. Indeed, the population of other developed nations, such as Germany and Hungary, is declining.

Europe and the former Soviet Union are presently the home of approximately 805 million people, or 14 percent of the world's population. It is projected that by the year 2025 these regions will have nearly 886 million people, or approximately 10.5 percent of the world's population. The United States and Canada also expect to increase in population from 292 million to 360 million by the year 2025. At the same time, they will also decrease in terms of percentage; presently these two North American nations make up 5.1 percent of the world population, but by the year 2025, they will make up only 4.3 percent.

The growth rate of the human population is much greater in the developing regions of the world, as Figure 4.7 shows. Today, there are 5.8 billion people inhabiting the world; 59.2 percent live in Asia, 12.9 percent in Africa, and 7.8 percent in Latin America. Together the underdeveloped regions of the world are home to about 4.6 billion people, or 80 percent of the world's population. These developing nations have a high momentum for population growth, due to the high proportion of individuals that have not reached childbearing age. In these regions, about 40 percent of the inhabitants are under the age of fifteen, whereas in the developed nations, the corresponding proportion is approximately 15 percent. This means that even if it were possible within the next few decades to obtain a fertility rate that would merely replace the parental generation, the population would continue to increase for fifty to seventy years thereafter.

Population growth rates are the highest in most of Asia, Africa, and Latin America due to the high degree of fertility and the dramatic decrease in mortality following World War II. Improved living conditions such as potable water, better nutrition, and medical services have significantly decreased infant mortality rates. This allows for many more children to reach sexual maturity and have families of their own. Unlike the developed nations of the world, most developing nations are not proceeding smoothly through the demographic transition. The developing regions do not have a major decline in the birth rate and appear to be stuck at the stage of low death rates and high birth rates.

In most of Asia, Africa, and Latin America, people have not yet enjoyed an increase in their living standard that the industrial revolution has brought to developed nations. In fact, the industrial revolution has barely penetrated some of these continents where many nations lack the capital to increase their level of industrialization. As Jodi Jacobson, a senior researcher at the Worldwatch Institute, suggests, "Slower economic growth in developing countries plagued by debt, dwindling exports, and environmental degradation means that governments can no longer rely on socioeconomic gains to help reduce births."[6]

In developing nations, offspring are considered as a form of "social security" system. Children may be the only source of security (income) for parents in their old age. In agricultural regions, children are a significant source of labor and, consequently, wealth. Lacking the very necessities of life, children may be the parents' source of wealth and pride. Culturally, there is often prestige for parents to have many children. A lack of education and birth control devices aggravates the impact of extreme poverty. Thus, birth rates remain high while death rates have been decreasing significantly due to advances in science and medical technology.

Of the projected population of the world by the year 2025, totaling about 8.47 billion, 7.2 billion or 84.4 percent will reside in Asia, Africa, and Latin

America. Hence, most of the increase in world population is expected to occur in developing nations. Asia will continue to have the largest number of people (4.9 billion or 57.8 percent), followed by Africa (1.56 billion or 18.7 percent), and Latin America (651 million or 7.7 percent). Currently, the high growth rate in Africa has been stable, whereas both Asia and Latin America show a declining trend.

There have been some exceptions to this rule. In some developing regions, such as China, the State of Kerala in India, and Sri Lanka, the population growth has been significantly decreased without a concomitant increase in per capita production. This decrease has been largely attributed to social policies that provide insurance against poverty and to the wide distribution of access to education, health care, and a means of subsistence. As noted earlier, the observed decrease in birth rates will not translate into slower population increase until the next generation. Populations of these developing regions will continue to mushroom due to the large percentage of their population being under the age of fifteen, which means that these individuals are just reaching the age of sexual maturity.

WORLD POPULATION POLICIES AND PROGRAMS

International agreements concerning world population policies have been quite controversial. Humans are the only species that have shown an ever-increasing rate of population growth; other species do not show this tremendous population increase due to regulation by the environment. We have noted that, with the development of modern medicine, death rates in most nations dropped dramatically. Continually high birth rates, along with the declining death rates, have caused the population to balloon in many poor countries.

Unlike wild animals, which do not have the intelligence to foresee signs of environmental resistance, humans have the ability to implement self-regulation. The growth attitude may have served its purpose of compensating for many humans who have died during earthquakes, pestilences, famines, and wars. To date, however, these factors have been reduced and no longer count for the large toll. Strategies for voluntarily reducing population size include increased mortality or reduced natality. The former method was practiced by some ancient civilizations, such as the Spartans, who let many infants die. The latter system is more humane and more easily accomplished through a host of birth control methods such as voluntary sterilization, birth control pills, diaphragms, contraceptive sponges, condoms, intrauterine devices, and the "morning after" pill. However, cultural and political factors have limited the access to and the use of birth control methods, even in some Western societies.

Voluntary and involuntary methods are either currently being used or cautiously being considered by many countries. In a voluntary population education program, the decision to procreate is left to individual couples. The government may play a role by discouraging couples from reproducing when the nation is overpopulated and encouraging reproduction when it is underpopulated. (All developed nations, excluding the United States, have a system of family allowances that provide payment to parents for supporting their children.) Nevertheless, the final decision will be made by the couple who is considering having children. In such a case, the population control program would be informative and advisory, but not mandatory. Such programs respect cultural and religious values, and are in full agreement with the United Nations Universal Declaration of Human Rights, which states:

Article 16 (3). The family is the natural and fundamental group unit of society and is entitled to protection by society and the State. . . .

Article 22. Some reports disclose coercive practices which have serious consequences for women, such as forced pregnancies, abortions or sterilization. Decisions to have children or not, while preferably made in consultation with spouse or partner, must not nevertheless be limited by spouse, parent or Government.

Most scientists agree that voluntary birth control programs offer the best prospects for achieving zero population growth and should be the primary method for doing so. Because some nations are already overpopulated, other measures of reducing growth have been undertaken. Concerned individuals in several developing countries have tried to grapple with the overpopulation issue even before the publication of Ehrlich's *The Population Bomb* in 1968. A review of the population policies in India and China is useful in illustrating both the successes and difficulties of implementing family planning programs.

India's almost miraculous triumph over malnutrition, malaria, poor sanitation, and medical illiteracy during the 1950s started it on a demographic roller coaster. The population of India grew exponentially from 360 million in 1951 to 439 million in 1961. By the late 1960s, it had increased to 520 million. India's efforts to expand agriculture could not keep pace with its population growth, and per capita food consumption actually declined from 12.8 ounces to 12.4 ounces per day. India's government became concerned that this gap would widen further, and it became the first nation to adopt an official family planning program. It did so during 1951–1952, although the program was not seriously pursued until 1965–1966. In the 1970s, supported by the World Bank and other Western organizations, India's government was the first to offer cash incentives or prizes (including portable radios) to men who already had fathered several children, if they accepted sterilization. Parts of India have employed deferred incentives in the form of old-age pensions, medical care, and economic rewards to be paid in the future to couples who have succeeded in maintaining small families. India's family planning program was a major contribution in reducing birth and fertility rates by approximately 50 percent. Nevertheless, the backlash by religious fundamentalists and individuals who believed that the government had no business regulating family size forced the government to relax its strict policy. In fact, the stringent policy was one of the major factors in the defeat of Indira Gandhi's government in 1977. Beginning in 1978, largely in response to the changing public attitudes, a new approach was initiated. India's government raised the age when marriage is allowed from eighteen to twenty-one years old for males, and from fifteen to eighteen years old for females. In the 1990s, the population growth rate began to increase again, and India's government debated on renewing its population control policy. As of this writing, India is the world's second most populated nation, with 935 million inhabitants. It is projected to have 1.4 billion people by the year 2025.

China, the largest developing nation with a population of 1.3 billion (21.5 percent of the world's population), presently has some of the most coercive population control policies anywhere. However, this is a sharp reversal of its earlier population policy. When the communists took over China, population growth was encouraged. In fact, Chairman Mao Zedong (Mao Tse-tung) stated in 1949, "It is a very good thing that China has a big population. Even if China's population multiplies many times, she is fully capable of finding a solution; the solution is production."[7]

This policy did not last for long. As infant mortality declined in the 1960s, it became apparent that a mushrooming population would strain even the most ambitious production plans. Overpopulation was leading to unemployment and shortages of housing and consumer goods, further hampering attempts to modernize the Chinese economy. The legal minimum age for marriage was raised to twenty for women and twenty-two for men. In addition, the government provides free birth control measures, as well as financial, educational, housing, and retirement incentives for couples who have only one child. The national government became so concerned with its overpopulation that it adopted a controversial one-child policy in 1978. In some regions of China, the government introduced such draconian measures as coercive abortion and sterilization. The population control policy also imposed penalties for having more than two children—those who refused to abort the third pregnancy were subject to fines. Between the 1960s and 1995, China's fertility rate dropped from 6 births per woman to about 1.9, lower than that of the United States. The main objective of the Chinese government is to achieve zero population growth by the year 2000. Yet, at the same time, some Chinese officials have expressed concerns that as the population of China is getting old, it may be necessary to encourage births again to supply needed workers to sustain China's economic growth. China's population is projected to surpass 1.5 billion by the year 2025 at the current growth rates.

China and India are not the only countries concerned about their expanding populations. In 1974, during the first international conference on population held in Bucharest, Romania, delegates from nations across the world had an opportunity to discuss issues of population and development and to prepare a population plan of action. At this conference, the developing nations were encouraged to initiate programs to help minimize the effects that overpopulation can place on economic development. At that time, very few developing nations had population policies or programs. There were concerns in the industrialized nations that the growth in population of the underdeveloped nations was the cause of the underdevelopment, and curbing such growth was a vital condition that had to be met if these nations were to become more developed.

Many developing nations resented the intrusion by developed nations in their sovereignty. They accused the developed nations of being selfish and argued that only a "New International Economic Order," where wealth would be redistributed from the rich nations to the poor nations, would solve their problems. Underlying this reasoning was the demographic theory that once a nation develops, its population stabilizes. Consequently, the advocates of aid to developing nations insisted that an equitable redistribution of wealth would lead to a decline in the population growth of the developing nations.

In spite of their different points of view, the nations gathered at Bucharest and adopted the "World Population Plan of Action" (WPPA). It served as a comprehensive model for national and international organizations in formulating and implementing international assistance in the field of population control policies.

Because many nations were either actively hostile or indifferent to population policies, the WPPA did not implicitly or explicitly set global targets for population stabilization. Rather, the Plan recommended that nations "consider adopting population policies, within the framework of socio-economic development." The role of international cooperation was stressed:

International co-operation, based on the peaceful co-existence of States having different social systems, should play a supportive role in achieving the roles of the Plan of Action.... Developed countries, and other countries able to assist, are urged to increase their assistance to developing countries ... and, together with international organizations, make that assistance available in accordance with the national priorities of receiving countries.[8]

The WPPA recommended that a comprehensive, thorough review and appraisal of progress made towards achieving the goals and recommendations of the plan should be undertaken every five years by the U.N. system. The first review and second appraisal of the plan were conducted in 1979 and 1984, respectively. The U.N. Economic and Social Council decided to convene in 1984 an International Conference on Population, under the auspices of the U.N. The purpose of the conference was

the discussion of selected issues of the highest priority, giving full recognition to the relationship between population and social and economic development, with the aim of contributing to the process of review and appraisal of the World Population Plan of Action and to its further implementation.[9]

Ten years after the Bucharest Conference, the U.N. International Conference on Population convened in Mexico City in August 1984. The purpose of this conference was (1) to evaluate the progress made by the implementation of the World Population Plan of Action; (2) to establish priority actions and objectives to expedite its implementation; and (3) to strengthen and sustain the momentum already produced by population control activities. At the Second World Population Conference, a more balanced perspective prevailed, and the notion that development itself was the best form of curbing population growth was accepted. Due to the comprehensive nature of the conference, and the intrusion of extraneous political issues, the conference achieved only limited goals. One of the chief political debates arose when the United States argued that population growth and the free market system should play a role in increasing economic development. At the 1984 conference, the United States, in stating its official policy, said:

Population and economic development policies are interrelated and mutually reinforcing. Based on historical experience, the twin objectives of economic growth and population stability without compulsion will most readily be achieved through adoption of market-oriented economic policies that encourage private investment and initiative. Such policies result in the most rapid increases in the standard of living, which in turn result in a lowering of birth rates as parents opt for smaller families.[10]

Moreover, the U.S. government expressed some concerns regarding coercion in order to achieve population control. The United States issued a policy statement that reaffirmed its long-standing commitment to voluntary family planning assistance programs for developing countries. Therefore, the United States refused to fund any organization or government that engaged in coercion to achieve its population goals.

Many delegates reluctantly affirmed the U.S. position and agreed that there was a link between population size and economic development in the improvement of the living conditions of the world's inhabitants. Achievements with respect to population and development required improvements in the other. All regions reaffirmed the major tenets of the World Population Plan of Action. In addition, it was restated that both individuals and couples have the right to make informed decisions about childbearing, and that it was important to respect cultural and religious values. Later, in 1993, the newly elected President of the United States, Bill Clinton, restored funding for the U.N. family planning programs.

From September 5 through 13, 1994, delegates from 183 nations met in Cairo to address global population issues once again. The circumstances had changed since the Mexico City Conference ten years earlier. The United States had a new president who was more sympathetic to supporting family planning programs. The world's human biomass had now increased by one billion. The emphasis was no longer solely based on demographic transition and economic growth. The delegates' discussion had a broader mandate, and took into consideration the connections among environmental degradation, demographic factors, and sustainable development. Four key goals were identified that could ameliorate the quality of life, as well as reduce fertility: (1) make contraceptives and family planning information and services universally available; (2) reduce infant and child mortality; (3) expand school enrollment for females; and (4) alleviate poverty.

Arguments between various religious groups and other delegates over the draft document's emphasis on a woman's right to have a safe abortion and empowerment received considerable attention from the media. Indeed, some Islamic countries, such as Lebanon, Saudi Arabia, and the Sudan, recalled their delegations in protest. The Roman Catholic church, which opposes birth control, insisted that a discussion on population control should be omitted. However, moderate delegates were very supportive of the overall statements and position of the Program of Action.

Financing was a key issue at the conference. The Program of Action estimated that $17 billion was needed to fund the family planning programs and the reproductive health services by the year 2000. The delegates also affirmed the importance of voluntary birth control methods.

As in the previous population conferences, no legally binding agreement was made. Although these conferences did not achieve the goal of world population stabilization, they were, nevertheless, instrumental in broadcasting issues regarding reproductive health, family planning, and population stabilization. They may even have helped to change attitudes on these major issues.

POPULATION POLICIES AND THE BILL OF RIGHTS OF THE UNITED STATES

Although Americans make up only 5 percent of the world's population, they are responsible for an incredible amount of overconsumption of resources and pollution problems. According to 1990 estimates, Americans consumed 35 percent of the Earth's raw materials and generated about 33 percent of the world's pollution. Consequently, an expanding United States population will have a profound negative effect on natural resources and even world environmental conditions.

Population Growth of the United States

The United States has one of the highest population growth rates compared to other developed countries (Table 4.1). Indeed, a large share of the estimated growth in the more developed nations will occur in the United States. The United States currently has 265 million inhabitants. This makes the United States the third most populated nation on the planet, after China and India. Its population is increasing by about two to three million people each year. According to the United Nations projection, the U.S. population will reach 322 million by the year 2025. As Figure 4.9 indicates, the population appears to be growing linearly. There are two important factors why the U.S. population growth has increased. The first can be attributed to the "baby boom" period from 1947 through 1964. During that period, having a larger family was more popular than today. This resulted in the "baby boom," a population increase whose momentum has not yet subsided. For example, in 1980 women who had reached their childbearing years had in-

Figure 4.9
Total United States Population: 1900 to Present and Projected to 2025

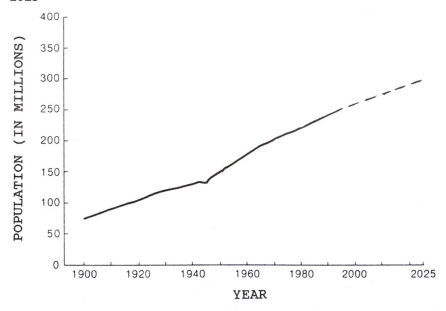

Projections are based on a birth rate of 2.1 children per woman.

Source: Population Reference Bureau. *World Population Data Sheet*. Washington, DC: Population Reference Bureau, 1995.

creased by more than a million, as compared to 1970. Even though these women were averaging 1.9 children each, because there were more of them, the total number of births grew from 3.1 million per year in the mid-1970s, to 3.6 million in 1980–1981. The second factor affecting the population increase in the United States was immigration. The immigration rate cannot be easily determined. Registered or legal immigrants number 500,000 per annum, but rough estimates of unregistered or illegal immigrants range from 100,000 to as high as 1,000,000 each year. (In a "normal" year, about 125,000 to 150,000 people leave the United States.) Both legal and illegal immigrants contribute to approximately 40–50 percent of the total annual population growth in the United States.

However, the growth rate has been declining since the 1950s (Figure 4.10); from 1980 through 1985 it rose to only about 1 percent. If current rates continue, zero population growth (ZPG), the point at which the population simply replaces itself, could be reached by the year 2030. Similar to other developed nations, the key reason for this decline is the number of

Figure 4.10
U.S. Population Growth Rates from 1960 to Present

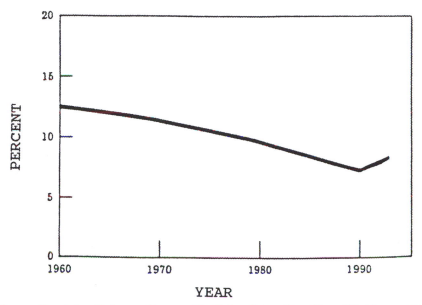

Source: Population Reference Bureau. *World Population Data Sheet*. Washington, DC:
Population Reference Bureau, 1995.

births. On the average, women now have only 1.8 children during their
childbearing years. In 1940 the average was 2.5 children, compared to the
late 1950s, which was 3.6 children. The replacement fertility level, or how
many infants are born to each woman during her reproductive years, that
would then result in ZPG is 2.1 children. Replacement level fertility is
slightly greater than two children because parents must replace themselves
and replace offspring who die before reaching sexual maturity.

Whether a population will endure depends on the relationship between
the following four variables:

Population change = (Birth + Immigration) - (Death + Emigration)

Inward movement (immigration) and outward movement (emigration)
do not affect the Earth's population change. They affect only local populations, or those in the political subdivision. As discussed, the major sources
of the U.S. population change are immigration and births. Therefore, in the
following paragraphs we focus on these two variables.

Immigration Policies of the United States

Historically, an expanding population has rarely been a concern of public decision makers. There is a notion that Americans pride themselves for being a nation of immigrants. In the past, due to a need for labor in industries and for people to settle in the western part of the country, there were few restrictions on immigration. People were also permitted to immigrate to the United States because of religious or political beliefs. But could a lack of resources or an increase in pollution cause a change in this policy?

Under the Constitution of the United States, Congress has the authority "to establish a uniform rule of naturalization." (Art. I, §8. Cl.4). The power of Congress over admission, exclusion, and deportation of aliens is absolute. Thus, an alien who seeks to enter the United States does so only upon such conditions prescribed by Congress. For example, Chinese immigrants were excluded to keep "inferior races" from entering the United States in 1882. In response to nativist forces, rather than economic factors, Congress passed the Immigration Act of 1924. The act developed a complicated system of country-of-origin quotas in order to retain the racial and ethnic composition of the U.S. population. In response to changing political, social, and economic conditions, Congress enacted the Immigration Act of 1965. This statute changed the priorities of the previous act from country-of-origin quotas to priorities based on family reunification, asylum for political refugees, and needed skills and professions.

During the 1980s and 1990s, immigration policies became matters of increasing public concern, especially where "downsizing" in employment and competition for social services were most intense. Advocates of the anti-immigrant policy, such as spokespersons for organized labor, nativists, and environmentalists, opened a broad-range challenge. Some argued that immigrants took jobs away from native-born blue-collar workers. New immigrants placed a heavy burden on the public health care system, education, housing, and general welfare needs. Some advocates of lower immigration levels contend that new immigrants, especially from the Third World countries, are profoundly affecting the common values held by Americans of European ancestry. Environmentalists in favor of zero population growth assert that the U.S. population should not increase, and if possible, should decline. The well-known environmentalist and professor of biology, Garrett Hardin, strongly supports a change in the U.S. immigration policy. Hardin not only holds that the policy would dampen population growth, but also feels that "an extravagantly multicultural nation is poorly positioned to compete with nations that have not succumbed to the siren call for more "diversity."[11]

Defenders of the present immigration policy, such as high-tech busi-nesspeople and cultural pluralists, pointed out the overall economic and long-term benefits of immigration. These groups assert that immigration has been the engine of U.S. economic growth, that most immigrants are hard working, and that the resurgence of regions with a high proportion of foreign born, such as Los Angeles, Chicago, and New York, owe their new economic vigor to the vitality that the new immigrants add to the commu-nity. In contrast, cities with a low proportion of immigrants, such as Pitts-burgh, Buffalo, and St. Louis, have shown markedly higher rates of unemployment. Cultural pluralists point out that a strict immigration policy would change the nature of America as we know it. A strict immigration policy would be contrary to the values held by many American citizens and the international community. Perhaps a more critical issue is whether America will continue to be envisioned as a world leader; a haven for the economically, religiously, and politically oppressed; a pluralistic nation distinguished by its quest for justice, fairness, and democratic ideals.

In response to the political activities of anti-immigration advocates, a number of changes in the laws affecting both legal and illegal immigrants have been made. The Immigration Reform and Control Act of 1986 con-tains provisions for civil and criminal penalties against employers who knowingly hire undocumented aliens. As of 1997, legal immigrants will no longer be eligible for many social-welfare programs that are available to citizens or nationals of the United States. Illegal aliens face strict scrutiny of documents, less opportunities to appeal if they are not granted refugee asy-lum, and more rapid deportation if they commit crimes. However, Congress has not reduced the annual quota of legal immigrants.

Birth Control Policies of the United States

U.S. public policy is unofficially pronatalist in that it encourages natality and births. The government has unofficially adopted certain policies and supported certain programs, such as an income tax structure that provides deductions for all children in a family and, until the Welfare Reform Act of 1996 capped categories of aid to dependent children and food stamps, in-creased allowances to welfare families for each child born. It is a capitalist system based on growth and a strong tradition that family size should be de-cided by the family rather than by the government.

Over twenty years ago, the Commission on Population Growth and the American Future (1972) suggested that "the nation welcome and plan for a stabilized population."[12] To reduce the high rate of unwanted births, the

Commission recommended the expansion of family planning and reproductive health services. However, because "birth rates in the United States
were declining and the issue was politically controversial," the recommendation was ignored.[13]

The rising environmental crisis has led to proposals to change the birth
control policies of the United States. Indeed, under the Clinton administration, the U.S. government shifted its policies toward supporting international family planning in countries with legalized abortion.

Birth control practices in the United States have been affected by medical technology, legislation, and court cases; additionally, the modern
women's movement has increased public pressure to maintain the "right" to
abortion, as defined by the landmark Supreme Court decision of *Roe v.
Wade* in 1973. On the contrary, various religious groups have mobilized politically against abortion. During the 1980s and 1990s, the politics of birth
control was one of the most powerful forces in American public life.

Public debate over birth control became more vocal as birth control became more available. Improved medical technology, especially, brought
more advantages to those wishing to limit their family size or increase their
fertility. The single most revolutionary medical advance in family planning
was "the pill." Since its approval in the 1960s as a birth control method, the
pill has slowly evolved and changed in response to research and societal
needs. To reduce estrogen-related side effects, the pill today contains less
estrogen and progestin than those used in the 1960s. In some cases, the
modern pills contain only progestin.

Furthermore, several lines of research offer better hope for alternatives in
controlling contraception. These biomedical technologies include a long-
term implant of a hormone that is placed under the skin to prevent the release of eggs from the ovaries, the use of prostandin vaginal suppositories to
stimulate a missed period, and a pill for males that reversibly inhibits sperm
production. These medical technologies will become powerful forces for
change.

Medical advances have improved the prospects of having children. They
include artificial insemination, which uses the semen of the husband or another donor to impregnate the female, and in-vitro fertilization, which uses
either both parents' gametes (egg or sperm cells) or donated gametes, im-
planting them in the uterus of either the biological mother or a surrogate
mother. Sperm banks permit an increase in the number of donors and theoretically make possible the selection of specific characteristics for babies.
Frozen embryos have increased the ease and success rates of these artificial

methods of reproduction. Some scientists predict that within ten years it will be possible to clone humans.

However, these medical advances have brought with them new social problems. According to public surveys taken in the 1980s and 1990s, many Americans worried that reproductive engineering could alter traditional family concepts and values. They fear giving power to reproductive engineers who would be in a position to manipulate our bodies, our genetic background, and the very fabric of our society. Concerns about the genetic manipulation of plants and animals raised public concerns about not only the "safety" of the food making possibility and population growth but also the control over human development. Advances in reproductive technologies have had legal, religious, and moral ramifications. These advances also have given people new choices, and sometimes forced them to make key decisions about increasing or decreasing their fertility. Legal questions have included: Would government intervention infringe on some liberty that is guaranteed by the Bill of Rights? And is the reproductive interest outweighed by the contrary interest of the government? More to the point, the capability of intervention, especially with regard to birth control, has given new choices to the government, potentially putting the government in the position to make decisions about reproduction that were previously beyond its control. In the political debates of the 1980s and 1990s, the question arose as to whether the government should, or if it even constitutionally had the right, to regulate such decisions in the name of ecological stability. (As described earlier, China and India have population policies that partly include involuntary birth control.)

Since the dawn of history, human beings have regulated their reproduction so that the number of children bears some relation to the parents' ability to support them. There are now a number of options, besides abstention, that are available to couples who wish to control their number of offspring. Each method offers certain advantages and disadvantages in its comparative applications, effectiveness, and side effects. The National Center for Health Statistics estimates that the percent of fourteen- to forty-four-year-olds who have ever used any method of contraceptive is 96.4 percent; female sterilization 23.1 percent; male sterilization 14.6 percent; pill 82.4 percent; implant 2.1 percent; injectable 4.5 percent; intrauterine device (IUD) 10.0 percent; diaphragm 15.2 percent; condom 82.2 percent; female condom 1.2 percent.[14] The use of some of these methods led to the U.S. Supreme Court's historic rulings. In the remaining part of this section we discuss the legal aspects of voluntary and involuntary use of contraceptives.

Voluntary Contraception. Contraceptive information and supplies were not always legally available. In 1873, in a crusade against what Congress deemed obscenity in literature, and in other forms of expression, the U.S. Congress enacted an antiobscenity statute known as the Comstock Act. This expansive obscenity legislation prohibited, in its omnibus regulation, the transportation in interstate commerce and mailing of "any article for the prevention of conception." States also had their own version of the Comstock Act that prohibited or restricted the sale, distribution, advertising, and display of contraceptive devices. On the other hand, all states by either statute or common law allowed the sale of all or some contraceptives by doctors, pharmacists, and licensed firms or individuals. Whatever the original reason of the Comstock Act and the states' regulation of contraceptives, the idea was to prevent contraceptive information and supplies from being readily available.

Almost a century later, in the court case of *Griswold v. Connecticut* (1965), the U.S. Supreme Court declared that laws that made it illegal for any person to use contraceptives, including married couples, were unconstitutional. The Court held that such laws infringed upon a constitutional "zone of marital privacy." Justice William Douglas delivered the opinion of the Court, and he noted:

Would we allow the police to search the sacred precincts of marital bedrooms for telltale signs of the use of contraceptives? The very idea is repulsive to the notions of privacy surrounding the marriage relationship.[15]

Six years later, in 1971, Congress repealed the Comstock Act.

In later cases, the Supreme Court made it clear that the *Griswold* case ultimately meant more than ensuring married couples access to contraceptives. The decision whether to use birth control was one of individual privacy; thus, the right belonged to singles as well as married couples (*Eisenstadt v. Baird* 1972). In the *Eisenstadt* case, the appellant had been convicted under a Massachusetts statute for giving an unmarried woman a package of vaginal foam at the close of a lecture on birth control that was delivered to a group of students at Boston University. The Court stated:

If the right of privacy means anything, it is the right of the individual, married or single, to be free from unwarranted governmental intrusion into matters so fundamentally affecting a person as the decision whether to bear or beget a child.[16]

Parents in the United States retain substantial rights concerning medical treatment administered to their young offspring. Unless there is a medical emergency, a parent's consent is required in order to render most types of medical treatment to a minor. Even piercing a child's ear or inserting an earring requires parental consent in some states. However, when it comes to reproductive autonomy, the Supreme Court has given minors more independence. The Supreme Court, while striking down a statute preventing the sale and distribution of contraceptives to minors under sixteen, stated that "read in light of its progeny, the teaching of *Griswold* is that the Constitution protects individual decisions in matters of childbearing from unjustified intrusions by the State."[17]

Terminating a pregnancy by induced abortion is one of the most widely practiced methods of birth control in the world. However, given a choice, women prefer not becoming pregnant in the first place to having an abortion. For both medical and humanitarian reasons, abortion is one of the most volatile issues of our time. The most controversial cases concerning voluntary contraception are *Roe v. Wade* and its progenies. Before we look at the implications of these cases, we should briefly study the biomedical aspects of abortion.

The removal of the embryo or fetus during pregnancy is known as abortion. The term is from the Latin abortus, "miscarriage." A large percentage of abortions are "natural," or spontaneous, and often occur very early in pregnancy. Abortion performed as a means of birth control is called induced abortion. Because this procedure is the termination of a pregnancy in progress, it is not considered a form of contraception. During the first trimester, abortion in the United States is usually performed in a hospital-associated clinic on an outpatient basis. With the use of light anesthesia, one of the safest and simplest methods involves dilating the cervix and gently drawing the embryo from the uterus using a suction device. The patient experiences only a strange pulling sensation. Another method is called dilation and currettage (D & C), in which the cervix is dilated and the lining of the uterus is scraped to remove the embryo and surrounding tissues. If these methods are to be used, the abortion should occur as early as possible, certainly during the first twelve weeks. Normally, complications from infection and excessive bleeding are minimal when the procedures are carried out by medically qualified personnel.

During the second trimester, abortion is more difficult. After this period, the placenta is firmly attached to the wall of the uterus, and such attempted abortions often result in extensive bleeding. Abortion during this time is generally done by injecting saline solutions into the amnion. Within hours after injection, expulsion of a nonliving or short-lived fetus follows. The woman will experience at least some labor pains just as if a normal birth were going to occur. If this method fails, the fetus must be removed by opening the uterus, a surgery known as hysterotomy. Occasionally, this operation is carried out through the vagina. However, most of the times it requires an abdominal operation similar to a cesarean section.

A study of the U.S. public health records indicates that deaths due to abortions have declined since it became legal in 1973. The death rate for women undergoing legal abortions during the first twelve weeks of pregnancy is about 3 per 100,000 women; for abortions undergone after twelve weeks, the number rises to about 25 per 100,000 women. Consequently, the earlier an abortion is performed, the safer it is for the woman. Overall, the present risk of dying from an induced abortion during the first eight weeks of pregnancy is now considered to be one-twentieth the risk of dying from pregnancy and childbirth. However, although not nearly as dangerous as full-term pregnancy, abortion is still more dangerous than most contraceptive techniques, such as the pill, sterilization, the IUD, and the like.

The abortion issue varies according to culture and time periods. In much of Asia and Africa, for example, infants traditionally had very few or no rights. The killing of an infant at birth has often been practiced or condoned. In Western societies, abortion before the stage of pregnancy at which the motions of the fetus first become perceptible (i.e., quickening) historically was not widely denounced or condemned. The politics of abortion did not emerge until the twentieth century. Under the common law, for example, women were free to have abortions at least until quickening. In 1972 a court in New York noted:

It is generally believed that abortion of a quick child was a high crime at common law, . . . although one commentator has argued persuasively that, in fact, it was not, that abortion was a purely ecclesiastical offense, punishable only by spiritual penalties and that the secular crime of abortion was created by the imagination of Sir Edward Coke who felt strongly that abortion after quickening should be punished and that the purely spiritual penalties of the ecclesiastical courts would not deter the people from it.[18]

Before quickening, a woman had up to one-third the chance of dying in childbirth, whereas abortions are less risky for the pregnant woman, so that abortions often saved women's lives. In fact, before the introduction of ether anesthesia (1846) and antisepsis (1867), any surgery was likely to cause death from shock or infection. Consequently, most religious and ethical opposition did not develop until after midwives and doctors found that washing their hands and clothes improved their patients' survival rates.

Once childbirth became less dangerous to a woman than abortion, attitudes began to change in the Western world. In some Western nations, anti-premature abortion laws were passed; and the Catholic Church, in 1869, equated early abortion with murder. Even in 1962, the Second Vatican Council stated that "from the moment of its conception, life must be guarded with the greatest care, while abortion and infanticide are horrible crimes."[19] Only indirect abortion is permitted, in order to save the life of the mother.

Due to newer methods of abortion described above, abortion to a woman is again less risky physically than bringing the fetus to full term. Once again, many governments began to encroach slightly on what had formerly been considered the inalienable rights of the fetus. Nations began to permit abortion for social, economic, or medical reasons. For example, in many countries, pregnancies resulting from rape can legitimize abortion.

During the early 1970s, in over two-thirds of the United States, abortion was a crime except when used to preserve the life of the mother; twelve states had changed their abortion statutes to be consistent with the American Law Institute Model Penal Code, which prohibited abortion except in cases where the pregnant woman's life or her mental or physical health was in danger, in the cases where the birth of defective offspring was very likely, or in cases of rape or incest. In 1970 abortion laws in Alaska, Hawaii, and New York were liberalized by legislation, and in the state of Washington by popular referendum. During the early 1970s, abortion was being reviewed in the courts in over half the states. At its 1972 meeting, the House of Delegates of the American Bar Association modified its opposition to abortion and approved a Uniform Abortion Act, which stated that abortion may be performed by a duly licensed physician upon request.

In 1973 the U.S. Supreme Court, in the landmark cases of *Roe v. Wade* and *Doe v. Bolton*, resolved the controversy arising out of the abortion statutes of Texas and Georgia. In the *Roe* case, a pregnant woman brought a class action, or an action on behalf of others in similar situations. She challenged the constitutionality of the Texas Penal Code that made it a crime to procure an abortion, as therein defined, or to attempt one, except with respect to an abortion procured or attempted by medical advice for the purpose of saving the life of the pregnant woman. Similar statutes were in effect in the majority of states.

In the *Doe* case, an indigent, married, Georgia resident sought an abortion. In Georgia, abortion was considered a crime unless performed by a physician duly licensed, and unless based upon his best clinical judgment that continued pregnancy would endanger a pregnant woman's life or injure her health, or unless the fetus would very likely be born with serious defects or unless the pregnancy resulted from forcible or statutory rape.

In addition to a judgment requiring that the mother be a state resident, the Georgia statutes also required the reduction to writing of the performing physician's medical judgment that an abortion is justified for one or more of the reasons just specified and advance approval by an abortion committee of not less than three members of the hospital staff. The Georgia statute was typical of the more liberal approach in existence in only about one-third of the states.

The Supreme Court held that the word "person" as used in section one of the Fourteenth Amendment does not include the unborn. The *Roe* case restrained the power of the states to forbid abortion; the *Doe* case held that states cannot create undue procedural requirements before abortion. For example, abortion statutes cannot require advance approval by a hospital abortion committee, because this procedure unduly restricts the patient's rights and needs that already have been medically delineated and substantiated by her personal physician. The aforementioned cases therefore render most states' laws on abortion illegal in whole or in part. Justice Harry Blackmun delivered the opinion of the Court:

With respect to the State's important and legitimate interest in the health of the mother, the "compelling" point, in the light of present medical knowledge, is at approximately the end of the first trimester. This is so because of the now-established medical fact . . . that until the end of the first trimester mortality in abortion may be less than mortality in normal childbirth. It follows that, from and after this point, a State may regulate the abortion procedure to the extent that the regulation reasonably relates to the preservation and protection of maternal health. . . .

This means, on the other hand, that, for the period of pregnancy prior to this "compelling" point, the attending physician, in consultation with his patient, is free to determine, without regulation by the State, that in his medical judgement the patient's pregnancy should be terminated. . . .

With respect to the State's important and legitimate interest in potential life, the "compelling" point is at viability. This is so because the fetus then presumably has the capability of meaningful life outside the mother's womb. State regulation protective of fetal life after viability thus far has both logical and biological justifications. If the State is interested in protecting fetal life after viability, it may go so far as to proscribe abortion during that period, except when it is necessary to preserve the life or health of the mother.[20]

The validity of the rationale and result of the *Roe* and *Doe* cases is being questioned in more recent Supreme Court cases. In *Webster v. Reproductive Health Services* (1989), the Supreme Court allowed the state to impose extra costs and burden on the pregnant woman if this would improve the fetus's chance of being born alive.[21] This requirement might include expensive tests and medical risk on the pregnant woman to be performed on any fetus over five months old for the purpose of determining whether the fetus is viable. Post-*Webster* abortion cases suggest that the U.S. Supreme Court will give the states a much greater right to regulate abortion. For instance, in *Hodgson v. Minnesota* (1990), the Court held that a provision of a Minnesota abortion statute, which required two-parent notification of a minor's abortion decision unless the minor obtains a judicial bypass, is constitutional.[22]

Further advances in birth control by abortion are on the way. Presently, the so-called "morning-after pill" uses hormonal inhibition by implantation and the administration of massive doses of an artificial estrogen for several consecutive days after intercourse. In the future, will the abortion issue disappear if improvements in birth control technology are developed? Or will the issue intensify and become more complicated as medical technology and practice increase the chances of "premature" babies and fetuses with defects to survive or have their defects treated while still in the womb?

Involuntary Contraception. The attitude of Americans toward forcing individuals to use contraceptives has dramatically changed during the last century. This change partially reflects the discovery of more-effective contraceptive methods, the scientific understanding of gene inheritance, and the abhorrent purposes for which sterilization was used by the Nazis and others.

Before the late nineteenth century, the only method of sterilization available was castration. This technique changes the secondary sex characteristics of the male, leaving castrated males, depending on their stage and physical condition at the time of castration, with high-pitched voices, little or no facial hair, reduced musculature, and, in many cases, the loss of any attraction to females. These castrated males are called eunuchs, and in the past were often placed in charge of harems or employed as attendants in a palace. Because it is a radical procedure with brutal side effects, castration has not been used as a sterilization method in the United States. In the 1890s, vasectomy was developed. This technique does not prevent the production of sperm or the secretion of hormones. Therefore, secondary sex characteristics are not affected. The sperm is simply reabsorbed, and the hormones are still excreted by the testes. The male can have intercourse and an orgasm and ejaculate a seminal fluid that does not contain sperm. The surgery can be done with local anesthesia in a physician's office, and new developments in the procedure offer the possibility of substantially increasing the probability of reversal.

In females, sterilization involves salpingectomy, the removal or tying of a section of each of the two fallopian tubes. When performed immediately following cesarean delivery or other abdominal surgery, salpingectomy is called tubal ligation. After the operation, the ovaries continue to function normally, but the sperm cannot reach the egg. The surgery does not alter the secondary characteristics of the woman; consequently it has no adverse effects on femininity. Salpingectomy is a more radical and costly procedure than a vasectomy, and it was not developed until 1910–1920. Originally, the salpingectomy was considered a major surgery and irreversible. Today, this procedure can be performed under local anesthesia and can be reversed.

Even before legislation was passed authorizing involuntary sterilization, some individuals were sterilized against their will. Dr. Harry Sharp was reported to have sterilized between six hundred and seven hundred boys in the Indiana Reformatory before 1907. However, later that year, Indiana became the first state to enact compulsory sterilization legislation. The statute authorized sterilization of prison inmates. By the end of the 1930s, thirty-two states had followed Indiana's example and enacted sterilization laws. Racial minorities and females were disproportionately subjected to such draconian measures. The laws often had a eugenic objective, intended to prevent the convicts from reproducing and thus transmitting their supposedly "antisocial genes" to future generations. (The term *eugenic* means "well-born," and eugenics is defined as the "science of improving the human race by better breeding.") The eugenics movement in the United States enjoyed significant growth in the early twentieth century, until the horrors of Nazi racism demonstrated the consequences of eugenic "logic." At that time, it was erroneously believed that criminal behavior was inherited genetically. Today, most scientists believe that behavior is not solely genetically controlled. They believe that the environment also plays a significant role in the development of personality traits and social behavior.

One of the primary legal issues presented by involuntary sterilization was that it was performed on retarded and epileptic patients in state institutions only. Those in private institutions were excluded from treatment. Therefore, public recipients were denied equal protection of the laws guaranteed by the Fourteenth Amendment.

Another issue presented by involuntary sterilization was that the states were in effect denying the right of procreation to some citizens not because they had violated the law, but simply because they were born "genetically defective." In 1913 a New Jersey Supreme Court judge questioned if such laws were constitutional: "Racial differences, for instance, might afford a basis for such an opinion in communities where the question is unfortunately a permanent and paranoid issue."[23]

States' high courts reacted differently to involuntary sterilization laws. The New York and Michigan courts declared that involuntary sterilization violated the equal protection clause. In 1918 the New York court added that the law was "inhuman in its nature."[24] In 1923 Michigan legislators enacted a more encompassing law by altering the original law to include sterilizations for "defectives" in both public and private institutions. In 1925 the new law was appealed, and the Michigan Supreme Court ruled that it was constitutional.[25]

In 1927 an eugenic sterilization case was finally reviewed by the U.S. Supreme Court. In that case, Carrie Buck was being forced to undergo a salpingectomy by the Superintendent of Virginia's Colony of Epileptics and Feeble Minded. Buck was an eighteen-year-old feebleminded inmate whose mother and illegitimate daughter were also similarly afflicted and confined to the same public institution.

In Buck's defense, it was argued that the salpingectomy was unlawful in that it neglected her constitutional right to bodily integrity. The statute denied Buck and other inmates of the state colony equal protection of the laws guaranteed by the Fourteenth Amendment. If the act was upheld to be constitutional, "we shall have set up Plato's Republic."

The Supreme Court declared that this was not a denial of equal protection of the laws. Justice Oliver Wendell Holmes delivered the often-quoted opinion of the Court:

We have seen more than once that the public welfare may call upon the best citizens for their lives. It would be strange if it could not call upon those who already sap the strength of the State for these lesser sacrifices, often not felt to be such by those concerned, in order to prevent our being swamped with incompetence. It is better for all the world, if instead of waiting to execute degenerate offspring for crime, or to let them starve for their imbecility, society can prevent those who are manifestly unfit from continuing their kind. The principle that sustains compulsory vaccination is broad enough to cover cutting the Fallopian tubes. . . . Three generations of imbeciles are enough.[26]

Carrie Buck left the institution after she was sterilized and lived with her husband for twenty-four years, until his death. One commentator appropriately remarked:

Through [Buck's] adult life she regularly displayed intelligence and kindness. . . . She was an avid reader, and even in her last weeks [she lived to the age of 76] was able to converse lucidly, recalling events from her childhood.[27]

Carrie's other daughter, who was one month old when the infant was judged "feebleminded" by a nurse, was listed on the honor roll in her two years of school. Unfortunately, the little girl died of measles at age eight. Had the Court known these facts, perhaps the decision in *Buck* would have been different.

During the 1930s, many Americans turned against involuntary sterilization after they learned that many inmates had been sterilized for reasons other than hereditary afflictions such as drunkenness, theft, and rape. Knowledge about Hitler's "final solution" for Jews and other "undesirables," which came to light during the 1940s, hastened the change in attitudes toward involuntary sterilization. During the 1930s, approximately 2,000 individuals were sterilized annually in the United States, but by the 1990s, less than 500 were sterilized annually. Most of those sterilizations were done for reasons other than eugenics, but racial and sexual bias still persisted in decisions about who would be sterilized. For example, the North Carolina Eugenic Board reported that between 1960 and 1968 the state had sterilized 1,620 individuals—97.7 percent of whom were women and 63.2 percent of whom were black.

In 1996, fourteen states still had laws authorizing sterilization of mentally disordered people (Alabama, California, Connecticut, Delaware, Iowa, Maine, Mississippi, North Carolina, Oklahoma, Oregon, South Carolina, Utah, Virginia, and Wisconsin). These laws range in scope from the sterilization of the feebleminded or mentally deficient, which was the case in all states, to the sterilization for "degeneracy" in one state (Iowa Code Ann § 145.9, 1972). In California, a law provides for the sterilization of inmates exhibiting "marked departures from the normal mental mentality."

Some of these statutes are based on the notion that the traits of the person intended to be sterilized are inherited and, hence, may be transmitted to future offspring. There are strong arguments that current genetic knowledge casts ominous clouds on many of the classifications found within the statutes.

Would the Supreme Court invalidate a statute that would force people who are not mentally disordered to practice birth control? The Supreme Court has used substantive due process as a way of protecting certain fundamental rights not expressly mentioned in the Constitution. To further illustrate this point, the right of privacy is not indicated anywhere in the Bill of Rights. Nevertheless, the Court has stated that the right of personal privacy is constitutionally protected. This right that relates to population control includes the following: the right to marriage, procreation, and travel. Under American constitutional law, when fundamental rights are infringed upon by the government, the laws are unconstitutional unless found to be necessary (narrowly drawn) to a compelling government objective. Given the Supreme Court's acceptance of the fundamental significance of privacy, not to mention the powerful cultural and political forces opposing federal intrusion into private affairs or public support for birth control, it is unimaginable that any government in the United States would be permitted to use involuntary birth control for the general population, even if such a policy were proposed. Ruling out involuntary contraception as a legal or moral option in the United States to control population, the government must seek alternative strategies.

Regarding environmental policy, strategic mechanisms that the government can use to see that its policies are adopted include some combination of (1) generalized pressure on individuals and industries, (2) direct regulation or strict enforcement of environmental standards, or (3) marketlike approaches that affect the supply and demand curves of environmental services. At the same time, the government can also (4) subsidize companies and individuals to encourage them to meet environmental standards, and (5) the government can control the integrity of the ecosphere where keystone natural resources are involved by regulating the production of these resources. All the preceding options may appropriately be used to regulate economic and social activities affecting environmental quality without infringing on fundamental rights. Moreover, there is no direct (one-to-one) correlation between population and pollution. Consequently, as the next chapter elaborates, the environmental issue is whether or not a country's population has exceeded its carrying capacity or sustainability.

NOTES

1. William Paddock and Paul Paddock, *Famine—1975! America's Decision: Who Will Survive?* (Boston: Little, Brown & Co., 1967).

2. Paul Ehrlich, *The Population Bomb* (New York: Ballantine, 1968), 7.

3. Cited in Stanley Johnson, *World Population—Turning the Tide* (Norwell, MA: Kluwer, 1994), 2.

4. Otis D. Duncan et al., "Marital Fertility and Family Size Orientation," *Demography* 2 (1965): 508–15.

5. Lester R. Brown, *World Population Trends: Signs of Hope, Signs of Stress* (Washington, DC: Worldwatch Institute, 1976); World Bank, *World Development Report* (Washington, DC: World Bank, 1984).

6. Jodi L. Jacobson, "Planning the Global Family," in Lester R. Brown et al., eds., *State of the World 1988* (New York: W. W. Norton, 1988), 151.

7. Cited in Smil Vaclav, *China's Environmental Crisis: An Inquiry into the Limits of National Development* (New York: M. E. Sharpe, 1993).

8. World Population Plan of Action, articles 100 and 104.

9. United Nations, *Review and Appraisal of the World Population Plan of Action* (New York: United Nations, 1986), iii.

10. Council on Environmental Quality, *The Fifteenth Annual Report* (Washington, DC: Superintendent of Documents, 1984), 483.

11. Garrett Hardin, *Living Within Limits: Ecology, Economics, and Population Taboos* (New York: Oxford University Press, 1993), 290.

12. Commission on Population Growth, *Population and the American Future* (Washington, DC: Superintendent of Documents, 1972), 110.

13. Michael E. Kraft and Norman J. Vig, "Environmental Policy from the 1970s to the 1990s: Continuity and Change," in Norman J. Vig and Michael E. Kraft, eds., *Environmental Policy in the 1990s* (Washington, DC: Congressional Quarterly, 1994), 13.

14. National Center for Health Statistics, *Vital and Health Statistics* (Hyattsville, MA: U.S. Department of Health and Human Services, 1997).

15. *Griswold v. Connecticut*, 381 US 479, p. 485–86 (1965).

16. *Eisenstadt v. Baird*, 405 US 438, p. 453 (1972).

17. *Carey v. Population Services International*, 431 US 678 (1977).

18. *Byrn v. N.Y.C. Health and Hosp. Corp.* 167 *New York Law Journal* No. 39, p. 5 (N.Y. App. Div., 1972).

19. Walter M. Abbott, ed., *The Documents of Vatican II*, 1966.

20. *Roe v. Wade*, 410 US 113 (1973).

21. *Webster v. Reproductive Health Services*, 109 *S.Ct.* 3040 (1989).

22. *Hodgson v. Minnesota*, 110 *S.Ct.* 2926 (1990).

23. *Smith v. Board of Examiners*, 85 *N.J.L.* 46, 88, p. 966. A. 963, 1913.

24. In re Thomson, 103 *Misc. Rep.* 23, 169 *N.Y.S.* 638 (1913).

25. *Smith v. Command*, 231 *Mich.* 409, 204 *N.W.* 140 (1925).

26. *Buck v. Bell*, 247 US 200 (1927).

27. Paul A. Lombardo, "Three Generations, No Imbeciles: New Light on *Buck v. Bell*," *New York University Law Review* 60 (1985): 61.

5

The Concept of a Self-Sustainable System

Earth Day in 1970 brought forth a new era in environmental consciousness. Over the last several decades, humans have made significant improvements in restoring some of the Earth's fragile environment. In the United States, the return of pollution-sensitive fish to many freshwater ecosystems, the cleaner air over some urban areas, and the rise from near extinction of the California gray whale and bald eagle are several examples of environmental "progress" since the first Earth Day. However, the world's environmental record leaves ample room for improvement. As indicated in Chapter 2, approximately half of all species are likely to become extinct by the end of the twenty-first century. The principal underlying causes of extinction are the expansion of the human population, trophy hunting, economic harvesting, deforestation, wetland drainage, urbanization, agricultural clearing, and pollutants. Humans are changing the environment and destroying natural habitats too rapidly for most species to adapt.

In addition, environmental disasters, such as the Bhopal gas leak in India (1984), the discovery of a growing hole in the stratospheric ozone layer (1985), the explosion of the Chernobyl nuclear reactor in the former Soviet Union (1986), and the predicted global warming during the twenty-first century, have forced people to recognize that human actions have destructive influences on the environment. All the while, the "four horsemen of the apocalypse" (pestilence, war, famine, and death) run rampant in much of the world. In 1996, one out of every three people worldwide suffered from either hunger, malnutrition, homelessness, or poor health care. Over the last half of the twentieth century, in any decade, there have been ten to thirty wars in progress, and ecological factors have played a key role in virtually all of them. The ecologist Charles Southwick concisely explained the connection between ecology and war:

What does war have to do with ecology? Revolution and war have always been considered within the realms of history, political science, sociology, economics, international relations, and perhaps psychology, but usually not in the domain of ecology. Ecology and war, however, are closely related in at least two ways. First of all, environmental factors—competition, space, resource requirements, and social interactions—are important causes of war. Nations compete for land, resources, power, and domination. Secondly, wars impact the environment directly, ravaging the Earth, often with longlasting effect, and indirectly, by diverting, consuming, and destroying valuable resources that could be used for productive purposes.[1]

If there is a lesson in the study of the environmental crisis culled from earlier chapters, it is that the crisis is complex and pervasive. There are connections between vanishing wilderness, pollution, overpopulation, poverty, war, and justice. During the fourth and final stage, some environmentalists see these connections and seek large-scale solutions that involve many aspects of society. The new environmentalists focus on the fate and condition of planet Earth as a whole.

In the 1980s, a "sustainable system" movement emerged to try to deal with the different aspects of the environmental crisis. In many aspects, this movement is quite different from previous ones. Indeed, "It may be the most important dimension of environmentalism because it implies a thoroughgoing transformation of industrial society."[2] Unlike wildlife conservation, antipollution, and the zero population growth movements of the past, which emphasized specific problems, the sustainability movement seeks to build a self-sustainable human system. This means meeting the needs of existing people without reducing the quality of the environment for future generations.

One valuable book that has helped society become aware of the concept of sustainability is *The Limits to Growth*, which was published in 1972 by a group of scientists from the Massachusetts Institute of Technology.[3] This group of researchers, coordinated by Donnella Meadows, developed computer models that connected global factors such as population growth, resource depletion, food shortage, pollution, and industrial output into one unifying model of world environmental conditions reaching into the twenty-first century. The computer projections suggested that if society did not significantly change its behavior, it would go beyond the Earth's carrying capacity with perilous economic and political consequences.

Since the publication of *The Limits to Growth*, Meadows and other scientists have developed better and more-sophisticated models that have incorporated potential technical advances in society. Most of these models have come to a similar apocalyptic conclusion: Infinite growth on a finite planet is impossible.

Understandably, the models proposed by these scientists lack the sophistication necessary to predict every aspect of human and environmental development. How, critics ask, can one be absolutely sure of future technical progress and the behavior of public decision makers? The models, nevertheless, have created a surge of interest among scientists and even policy-makers about the concept of sustainability.

The Organization of Petroleum Exporting Countries (OPEC) oil embargo of 1973 further energized the interest in sustainability. The resulting gasoline shortage and price increases, which stalled economic growth in much of the industrial world, clearly sparked a pessimistic prediction of a future era in which resources would become scarce.

Sustainability also leaped into prominence with photographs of the Earth taken from space. In the words of the United Nations Report on Environment and Development:

In the middle of the 20th century, we saw our planet from space for the first time. Historians may eventually find that this vision had a greater impact on thought than did the Copernican revolution of the 16th century, which upset the human self-image by revealing that the Earth is not the centre of the universe. From space, we see a small planet and fragile ball dominated not by human activity and edifice but by a pattern of clouds, oceans, greenery, and soils. Humanity's inability to fit its doings into that pattern is changing planetary systems, fundamentally.[4]

What really gave the sustainable movement a strong surge was the publication in 1987 of *Our Common Future* by the United Nations World Commission on Environment and Development. The commission, consisting of twenty-two eminent participants from both developed and developing nations, was charged with identifying international long-term environmental strategies. Chaired by Prime Minister Gro Harlen Brundland of Norway, the commission was highly influential in providing an intellectual framework of the global environmental crisis. *Our Common Future* defines sustainable development as that which "meets the needs of the present generation without compromising the ability of future generations to meet their own needs."[5] Sustainable development guarantees that future generations have access to the "social capital"—people and natural resources—to create a life that is at least equal to that of the present generation. The report proposed twelve key priorities necessary to "sustain human progress into the distant future":

1. Achieve population control.
2. Decrease poverty, inequality, and debt in developing countries.
3. Develop sustainable agriculture.
4. Protect the vanishing wilderness and genetic diversity.
5. Protect ocean and coastal resources.
6. Protect freshwater quality and improve water efficiency.
7. Improve energy efficiency.
8. Develop renewable energy resources.
9. Control greenhouse gases and other air pollutants.
10. Protect the stratospheric ozone layer.
11. Minimize wastes.
12. Reduce military spending so that money can be diverted to funding sustainable development.

The report provided an agenda for advocates of sustainability, and the imprimatur of the United Nations implied a worldwide obligation to act on the recommendations.

HUMAN SOCIETY AS A SELF-SUSTAINABLE SYSTEM

The concept of sustainability drew on an idea that had a history of over forty years. Known variously as "spaceship earth," "anthroposystem," "sustainable society," and "sustainable development" (see Table 5.1), these overlapping notions suggested that the time is ripe for the concept of sustainability to appear. Whatever term is used, they all support the idea of a society in balance with its surroundings, a self-sustainable system.

In discussing the concept and philosophy of "sustainability," it is easy to get lost in details about our vanishing wilderness, toxic substances, ozone depletion, global warming, acid rain, overpopulation, depletion of resources, steady-state economy, and hundreds of other concerns. Therefore, it is wise to have an intuitive model that can serve as a framework, to bring the self-sustainable issue back into perspective whenever it gets out of focus.

Over twenty years ago, Miguel Santos proposed the model of the anthroposystem to refer to "the orderly combination or arrangement of physical and biological environments for the purpose of maintaining human civilization."[6] An anthroposystem is a structural and functional unit of the environment; it can be considered a self-contained system, provided that it has an energy source. An anthroposystem requires a sustainable economy to recycle used products and resources, and a pollution-minimizing economy. Though the model shown in Figure 5.1 is certainly oversimplified, it is

Table 5.1
Changing Perception of a Stable Human-Environment System

As a Spaceship Earth Concept:

The closed earth of the future requires economic principles that are somewhat different from those of the open earth of the past. . . .The essential measure of the success of the economy is not production and consumption at all, but the nature, extent, quality, and complexity of the total capital stock, including the state of the human bodies and minds included in the system. In the spacemen economy, what we are primarily concerned with is stock maintenance.

Source: Kenneth E. F. Boulding, 1966. "The Economics of the Coming Spaceship Earth." In Henry Jarrett, ed., *Environmental Quality in a Growing Economy*. New York: Johns Hopkins University Press, 9.

As a Federal Law:

It is the continuing policy of the Federal Government . . . to use all practical means and measures . . . in a matter calculated to foster and promote the general welfare, to create and maintain conditions under which man and nature can exist in productive harmony, and fulfill the special, economic, and other requirements of present and future generations of Americans.

Source: National Environmental Policy Act of 1969, 42 U.S.C. s/s 4321.

As an Ecological Concept:

The anthroposystem is the orderly combination or arrangement of physical and biological environments for the purpose of maintaining human civilization . . . built by man to sustain his kind.

Source: Miguel A. Santos. *Ecology, Natural Resources and Pollution*. London: Living Books, 1975, p. 52.

As a Sustainable Society:

That sustainable society is one that lives within the self-perpetuating limits of its environment. That society . . . is not a "no-growth" society. It is, rather, a society that recognizes the limits of growth . . . a society that looks for alternative ways of growing.

Source: James Coomer, ed. *Quest for a Sustainable Society*. Oxford: Pergamon, 1981, p. 1.

As Development:

Development that is likely to achieve lasting satisfaction of human needs and improvement of the quality of life—by integrating conservation into the development process.

Source: Robert Allen, *How to Save the World.* London: Kogan Page, 1980, p. 23.

As an Economic Definition:
    The concept that current decisions should not impair the prospects for maintaining or improving future living standards. . . . This implies that our economic systems should be managed so that we live off the dividend of our resources, maintaining and improving the asset base.

Source: Robert Repetto. *World Enough and Time.* New Haven, CT: Yale University Press, 1986, p. 15–16.

As a United Nations Policy:
    Development which meets the needs of the present without compromising the ability of future generations to meet their own needs.

Source: World Commission on Environment and Development. *Our Common Future.* Oxford: Oxford University Press, 1987, p. 43.

Improving the quality of human life while living within the carrying capacity of supporting ecosystems.

Source: World Conservation Union. *Caring for the Earth.* Gland, Switzerland: UN Environment Programme, 1991, p. 10.

Figure 5.1
The Concept of Sustainability, Using the Anthroposystem Model

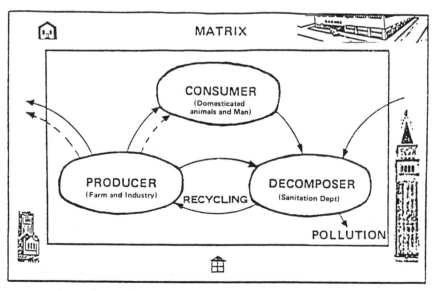

Source: Adapted from Miguel A. Santos. *Ecology, Natural Resources and Pollution.* London: Living Books/New York: Diplomatic Press, 1975.

useful in the strategic planning for building a sustainable society and it provides an intellectual guideline for linking wildlife conservation, pollution, and population issues.

We divide the anthroposystem into components, based upon how they influence humankind's chances of survival in a stable environment. At this point, the anthroposystem may be broken down into four major components, which can then be separately defined and logically analyzed. The four components of an anthroposystem as illustrated in Figure 5.1 are matrix, producers, consumers, and decomposers. The *matrix* is composed of all nonliving and nonproductive parts of the system—such as buildings, streets, land, air, and water. It also provides the structure or fabric in which the other components operate. The *producer* is the component that manufactures or yields products. There are three categories of producers: agricultural, industrial, and ecospheric resources. The agricultural producers are green plants, such as wheat, barley, rice, and corn. The industrial producers are the machines and tools utilized by humans to produce shelter, clothing, transportation, and so on. The ecospheric resources include parts of the ecosphere used by humans as a source of natural resources such as fish used for economic harvesting or forests as a source of oxygen. The *consumers* consist of humans and their domesticated animals. The *decomposers* in the anthroposystem are the waste-water treatment plants, resource recovery plants, electrostatic precipitators, spray collectors or scrubbers, and the natural decomposers of the ecosphere that can eliminate some by-products of society in a sustainable manner. A fully functional decomposer component also serves to maximize the recovery of resources. As late as the mid-1990s, however, resources that are not needed continue to be dumped into surrounding environments rather than be recycled. This dumping process causes contamination of the environment and depletion of valuable natural resources.

Because different human environments or societies grade almost undetectably into one another, the division of an anthroposystem is a convenient way to organize our thinking. There is, however, a level of human organization that is larger than the anthroposystem yet easier to see as a united whole. This is any nation, state, or country. But air, water, transboundary pollutants, other molecules and compounds do not stop at national boundaries. An anthroposystem exists in the ecosphere and should be viewed in an ecological context. The boundary between an anthroposystem and its surroundings is an imaginary one, used only for convenience in discussion. An anthroposystem is an artificial system produced by human efforts, and it exists as a result of these efforts.

Ecosystems (for example, ponds and forests) and anthroposystems are interwoven, but a separation between the two enables us to organize these concepts. For obvious reasons, the divisions are mentally constructed and designed in order to cope with the tremendous diversity of the environment. They are not laws of nature; the environment does not come in two convenient categories labeled ecosystem and anthroposystem. For example, the nitrogen element that forms a part of our eyes may have been a part of a tadpole's nervous system that was consumed by the fish we ate. Thus, there is an interrelationship of parts between and within ecosystems and anthroposystems. The real world is composed of a mosaic of interrelated natural ecosystems and artificial anthroposystems.

In these interrelated or coupled systems, there is a cause-and-effect relationship whereby different variables affect one another, as with the cogs in a machine. Turning one part inevitably causes motion somewhere else. The Council on Environmental Quality eloquently stated the complexity of the environment:

As our understanding of the physics, chemistry, and biology of the Earth has progressed, more and more evidence has accrued that the separate components of the planet are really not separate at all, but are interrelated in both obvious and complex ways. Ultimately, we cannot separate the geosphere (oceans, atmosphere, ice cover, and solid earth) from the biosphere (aquatic and terrestrial, including man), because events in one part of the Earth system are intimately related to happenings elsewhere. As a simple example of this linkage, consider rocks and rain reacting to form soils, while plants and animals through their mutual activities, in turn, influence the composition of both the soil and the atmosphere.[7]

In 1988 Edward Shevardnadze, foreign minister of the Soviet Union, viewed the world from a very different political perspective but reached a similar conclusion. In an address to the U.N. General Assembly, he stated that "the dividing lines of the bipolar ideological world are receding. The biosphere recognizes no division into blocs, alliances or systems. All share the same climatic system and no one is in a position to build his own isolated and independent line of environmental defense."[8] It is ironic that the Soviet bloc in which Shevardnadze was such a vital component has fragmented while the biosphere has remained intact. Indeed, from an ecological perspective, there is only one biosphere.

If an anthroposystem exceeds the point where its demand is greater than the sustainable yield of resources and the assimilative capacity of its matrix for pollutants, then it has exceeded its carrying capacity. This reduces production, thus triggering a change in the system's structure and function.

Consequently, the idea of carrying capacity focuses on the interaction between a society, its activities, and the surrounding matrix. It highlights natural thresholds that might otherwise remain unclear.

An ideal anthroposystem satisfies its needs without diminishing the prospects of surviving in space and time. Evaluated by this measure, contemporary society fails to meet this criterion. The decomposers in human systems are not as developed as in natural ecosystems. Ecosystems rely on their decomposers to break down dead plants and animals and to recover or recycle waste materials. Human systems recover very few of their waste materials, whether domestic, industrial, or agricultural, which are then dumped into the environment, where they accumulate and may cause pollution. The parasitic and destructive nature of human systems results in the depletion of natural resources and therefore the pollution of the environment.

The anthroposystem depends upon natural resources for aid and support. These resources are essential for the survival of the anthroposystem, and for determining the size and organization of society. However, as suggested by the sociologist William Catton, "With different organizations and technologies, one population of humans can be a very different sort of ecological entity than another human aggregate."[9] In effect, there would be a different level of carrying capacity depending on the degree of intellectual, moral, and personal values of each anthroposystem. In other words, the acceptable carrying capacity depends on what a particular society considers to be its wants and needs, without reducing the quality of life for future generations. The anthroposystem that chooses an American lifestyle must necessarily be smaller than one that chooses a communal lifestyle.

There are two categories of natural resources: functional and habitat. Functional resources include factors required for the biological needs of humankind and domesticated organisms and for industrial processes. They provide energy and/or the materials required for the metabolism of the system. Functional resources also include agricultural nutrients, solar energy, limestone, chromium, aluminum, iron, oxygen, and water, to name several. Habitat resources include factors not required for energy and/or materials—some examples include temperature and space. Other resources can serve as both functional and habitat resources. For example, solar energy warms the Earth's surface and drives winds and the hydrologic cycle; moreover, a small percentage of solar energy powers photosynthesis.

Resources can be divided into two categories: renewable or nonrenewable. Renewable resources can be made usable by resource recovery or be replenished by natural processes. Some examples of renewable resources are food crops, domestic animals, wildlife, forests, fresh air, fresh water, and fertile soil. Nonrenewable resources cannot be made new, nor can they be restored to their former condition. Mineral resources (such as iron and copper) and fossil fuels (gas, petroleum, and coal) are nonrenewable resources. Each removal from its origin is final and irrevocable. These resources either exist in a fixed supply or cannot be replaced as fast as they are used.

If the rate of consumption or loss of a given renewable resource exceeds the maximum replenishment or harvesting for an extensive period of time, the stocks will be depleted. The human system, dependent on the stocks, would be impoverished and eventually perish. The lesson here is simple but important: If the consumption rate in a population exceeds the replacement rate (no matter how small the difference), the relative scarcity will grow exponentially. Lester Brown elaborates on the effects of a growing population with its concomitant consumption of resources:

These spiraling human demands for resources are beginning to outgrow the capacity of the earth's natural systems. As this happens, the global economy is damaging the foundation on which it rests. Evidence of the damage to the earth's ecological infrastructure takes the form of collapsing fisheries, falling water tables, shrinking forests, eroding soils, dying lakes, crop-withering heat waves, and disappearing species.[10]

Resource recoveries include the productive use of energy and matter that would otherwise be disposed of as waste. Energy recovery refers to obtaining energy from organic wastes, as in refuse-derived fuel incinerators. Matter recovery includes recycling, reusing, and material conversion. The most common method of matter recovery is recycling, which is the reprocessing of wastes to recover the original raw material. For example, old aluminum cans are usually melted and recast into new cans and wood fiber is recovered from wastepapers. Reusing products differs because a product or good is reused in the same form (for example, cleaning glass bottles). In material conversion, the waste is used in different forms (like road-paving material from auto tires).

When material resources are recovered, this makes the human system a closed system for that particular resource because no significant exchange of resources occurs between that system and its surroundings. All wastes are completely treated so that there is no discharge of pollutants into the environment. In such a closed system, the resource can, theoretically, last indefinitely.

The real world differs significantly from the closed, ideal anthroposys-tem model. First, in every type of manufacturing, raw materials are ex-tracted, refined, processed, and transformed into finished products. Along the way, these processes generate wastes that are the inevitable by-products of industry. Second, based on the second law of thermodynamics, energy used is never completely efficient. This natural law states that whenever en-ergy is converted from one form to another, some potential energy is lost. For example, a car converts the chemical energy stored in gasoline to the ki-netic energy of movement. In the process, 75 percent of the energy is imme-diately lost as heat. The same thing occurs in organisms. As glucose is used as energy by organisms, only 40 percent of the energy is captured as bio-logically useful energy; the other 60 percent of the chemical energy is dissi-pated as heat. The consequence of the second law of thermodynamics is that energy cannot be recycled or reused. Therefore, in theory and in fact, real anthroposystems are open systems as far as energy is concerned. Third, as mentioned previously, the ecosphere is composed of a mosaic of interre-lated human and natural systems. Thus, human systems tend to be open sys-tems because resources are both imported and exported (Figure 5.2).

Figure 5.2
Human Societies—Open Systems

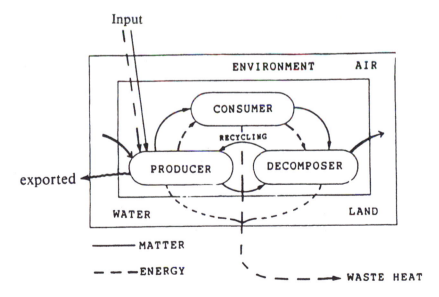

Source: Adapted from Miguel A. Santos. "Quantification of the Anthroposystem Concept." *Journal of Environmental Systems* 12 (1983): 351–61.

There are a number of drawbacks in the use of the anthroposystem model. The most obvious is that the complex industrialized, congested systems require more science and technology. The model assumes that humanity has the knowledge, understanding, foresight, and wisdom to guide the orderly and sequential development and ensure the environmental security of planet Earth. In 1974 Miguel Santos was skeptical as to whether or not society was "ready to take helm of the planet earth and guide its orderly and sequential development,"[11] but now he is convinced that we must somehow ensure the environmental security of planet Earth. It is the only way society can equitably deal with the looming extinction of nature, the destruction of the global commons, and the presence of pestilence, and in the process, achieve some sort of self-sustainable system.

The anthroposystemic worldview is in sharp contrast to that of some environmentalists (e.g., deep ecologists). Some members of the deep ecology movement believe in voluntary simplicity: being in touch with nature, back-to-the-land communalism, and the sacredness of nature as a method of achieving a sustainable future. On the other hand, the anthroposystem model relies heavily upon environmentally benign technological innovations and an understanding of how the ecosphere works as a way of achieving a sustainable future. In fact, under the anthroposystem model, it is understood that achieving a self-sustainable system is a governmental duty; science and technology are the means to fulfill these duties.

Some radical environmentalists have argued that science begets technology, and technology is usually the cause of more severe problems rather than a solution to them. For example, since the first successful test detonation of the atomic bomb in 1945, nuclear physicists have wrestled with the scientific interest in understanding the nature of nuclear energy and the development of nuclear energy for military purposes (atomic and hydrogen bombs). Meanwhile biologists have attempted to understand and manipulate the genetic properties of plants and animals and apply that science to an understanding of human development. During the 1980s and 1990s, this has led to nightmares of the possible creation of mass numbers of identical people (the cloning of soldiers). Other scientists, while sharing concerns about abuses of science and the danger of unquestioning acceptance of all scientific and technological developments as "good," point to the wide array of technology that modern societies use to correct environmental and medical conditions. To consider these materials and methods as being "bad," they argue, means also having to condemn false teeth, eyeglasses, contact lenses, birth control devices, hearing aids, vitamin pills, light bulbs, telephones, books, glass, waste-water treatment plants, clothing, and more.

Science and technology, they insist, are tools that may be used by mankind to abate needless human suffering and to create a functional environment, that is, an anthroposystem.

In the Western world, the central point of the technological debate has focused on how to decide, prudently, when technology should be put to use. In the twentieth century, modern culture has grown ambivalent about technology, welcoming the "progress" of new tools and processes while fearing the loss of control to machines. In popular culture, movies such as Charlie Chaplin's *Modern Times* (1936) and Stanley Kubrick's *2001: A Space Odyssey* (1968) warned that technology ("the machine") threatens to dull the human spirit and even rule humanity. Scientists point out that technology, per se, is neither "good" nor "bad," and thus consider it naive to interpret technology as inherently evil. How science and technology are deployed depends on the rational or irrational decisions of social and political policymakers, not on the technology itself.

Unfortunately, the unanticipated effects of technology can be frustrating. When the insecticidal properties of DDT were first discovered in 1939, no one imagined that it would one day show up in the tissues of birds. Lester Brown and Sandra Postel pointed out:

Corporations manufacturing the family of chemicals known as chlorofluorocarbons, for example, surely did not intend for these compounds to deplete the ozone layer. Their goal was to produce efficient refrigerants, a practical propellant for aerosol spray cans, and a chemical agent used to make foam products.[12]

One cannot help but to ponder over the thoughts of Socrates. He believed that it was better to be an unhappy person than a blissful pig. The pig is content with circumstances as they are as opposed to a human's dissatisfaction, hopes, or expectations. Freedom to take the risk of change and improvement is what makes us human. The best quality of humanism is courage, coupled with reasoning. Moreover, the surprising and potentially catastrophic effects of ozone depletion and global warming require that society anticipate the future—ignorance is not bliss, as far as environmental problems are concerned.

RELATIONSHIP BETWEEN THE STEADY-STATE ECONOMY AND SUSTAINABILITY

A concept similar to the self-sustainable system, though somehow narrower in scope, is the idea of a "steady-state economy." These two perspectives are not mutually exclusive; they both contribute to an understanding of a system that is life-sustaining instead of life-threatening. Each is valid and can be used to address different analytical issues. Let us examine in detail the concept of the steady-state economy.

There exists a contradiction between capitalism and the steady-state economy. Capitalism is based on growth—we expect the gross national product (GNP) to increase annually. To keep the growth rate from tapering, the government encourages a constantly expanding investment in business, a continuous flow of goods and services, and quantities of consumers to provide markets. The capitalist government contends that these criteria are essential in order to create jobs for the growing population, and necessary to keep incomes in line with inflation.

Traditionally, Western economists have advocated policies that encouraged an increase in economic growth. They contend that zero economic growth has undesirable income distribution effects, and that growth has been the means by which the economically disadvantaged raise their living standards. In this line of thinking, for example, the United States cannot aid developing nations by producing and consuming less; rather, it best can provide help to these nations by improving American productivity. An expanding economy can carry out new programs to minimize poverty and improve environmental quality. In other words, economic growth reduces the problem of scarcity of income. Underdeveloped nations cannot be assisted if the United States endorses a policy of zero economic growth or asks its citizens to share more of their existing resources.

Advocates of increasing growth also point out that it is unclear whether a rising GNP means an increase in net pollution. Ancient and medieval cities had their own special types of pollution, such as manure on unpaved roads and pathogens of typhoid fever in drinking water. Modern cities have other brands of pollutants, such as CFCs and asbestos. Modern society makes choices to produce things that are detrimental to the environment (e.g., air conditioners) or to produce things that do not pollute the environment nearly as much (e.g., solar heating devices). Consequently, these economists caution us not to confuse the control of growth with the control of pollution.

Advocates of increased growth believe that the planet's resources are not finite and can grow. In 1984 the economist Julian Simon, who put an incredible amount of faith in humankind's foresight and the rate of technological progress, elaborated on this point by arguing that the free play of market forces—including natural ones—will promote equilibrium. To illustrate, if fossil fuels are depleted, new substitutes will exist. Simon wrote:

Nor do we say that a better future happens automatically. It will happen because people—as individuals, as enterprises working for profit, as voluntary groups, as governmental agencies—will address problems and will probably overcome, as they have throughout history.[13]

Supporters of zero economic growth point out that a rise in per capita gross national product (GNP) is not the best indicator of economic well-being. Foremost among its problems is that the GNP is a barometer of the goods and services produced, rather than consumed, per year. Economists normally view the consumption of goods and services, and the pleasure derived from them, as the real measure of economic well-being. Because the GNP is a production-oriented indicator, it is apparently inadequate in this respect. Despite the fact that the salaries of paid domestic workers, who perform identical services, are incorporated, one shortcoming of the GNP as a welfare index is that it does not take into account the services of people who do their work at home: cooking, cleaning, and maintaining the household. The GNP is deficient when an employee chooses to work fewer hours in exchange for more leisure time even though such a decision, if voluntary, is to that employee's benefit. This deficient consideration of leisure is another defect of the GNP as an indicator of well-being.

The most environmentally relevant aspect of the GNP is that it does not include many of the waste products related to production of things that are measured by the GNP. Many of these by-products are detrimental to the health and welfare of society. The satisfaction that people derive from a healthy environment, or the time spent in a national park or wilderness area, cannot be indicated in the GNP. The damage to biota by pollution is similarly not taken into account. Actually, additional defensive expenditures required to fight the adverse damages of pollution would be reflected in an increase of the national account. For example, the *Exxon Valdez* oil spill off the Alaska coast in 1989 actually increased the GNP due to the fact that $2.2 billion was spent on labor and equipment to clean up the mess. Perhaps the most significant benefit of pollution control expenditures—improved human health and environmental stability—will never be known, because one can only guess what might have happened if there were no pollution control.

Another factor by which critics of growth economics warn that growth cannot continue is through the law of diminishing returns. For example, the mineral deposits and amount of arable land available to a country can be augmented only within narrow limits. As population increases, more and more laborers will work on a finite amount of arable land and output per worker will gradually decrease. Each worker is combined with a progressively smaller area of land, so productivity diminishes.

The history of Ireland in the nineteenth century, with land subdivided into smaller plots forcing greater reliance on the potato as its staple, bore grim witness to the diminished returns of overuse. In the twentieth century as well, the overgrazing and overcultivation of lands in developing nations in Africa, Asia, and Latin America were, in part, caused by excessive population pressures on limited land.

According to the law of diminishing returns, output per laborer should decrease continuously until a subsistence standard of living is approached. However, in theory, if each laborer is equipped with more capital and better technological know-how, the labor force itself becomes better and all resources become more efficiently managed. Under these conditions, the law of diminishing returns can be delayed, but only up to a point. It is unquestionable that Earth's resources are finite, but the problem facing environmentalists, economists, and public decision makers in the 1990s has been at what point will the law of diminishing returns begin to operate?

During the 1990s, the quantity of land utilized for crops on Earth increased, but at slower rates than ever before. Millions of hectares were brought under cultivation for the first time each year, but nearly as much was taken out of cultivation to be urbanized, returned to pastures or forests, or abandoned. Still, the overall impact on the land was profound. With the Earth's population growing exponentially and total arable land increasing minimally, arable land per individual is declining.

Agricultural production has grown more or less linearly despite the relatively small increases in arable land (see Figure 5.3). Africa and most of the republics of the former Soviet Union, however, experienced a decline in production during the 1990s. In fact, in Africa and the republics of the former Soviet Union, production per capita fell. According to a 1994 report by the World Resource Institute:

Agricultural production in much of the developing world has been an extraordinary success over the past several decades, but the pressure to grow more food will continue as populations rise. With land growing scarcer, most future production gains will have to come from greater average yields per hectare. Yet in Asia, yield gains are generally slowing, while in sub-Saharan Africa the gap between supply and demand is expected to widen.[14]

It seems that developing nations may be inadvertently adhering to the Malthusian doctrine. Recall from Chapter 4 that Thomas Malthus observed that population increase among the poor was exponential, whereas food-growing increase was, at best, an arithmetical progression. Consequently, the standard of living can never rise far above subsistence levels because of

Figure 5.3
Food Production and Index of Per Capita Food Production by Region

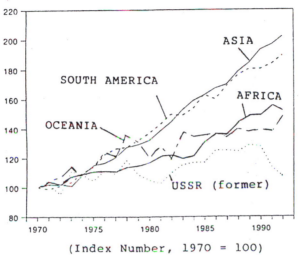

Index of Food Production by
Region, 1970–92

(Index Number, 1970 = 100)

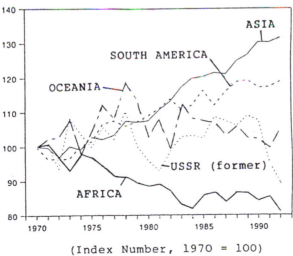

Index of Per Capita Food
Production by Region, 1970–92

(Index Number, 1970 = 100)

Source: U.N. Food and Agricultural Organization. *Agrostat PC*, on diskette. Rome: FAO, 1993.

constant population pressure on food supply. The results, he predicted, were inevitable—periodic famines, plagues, and wars.

Today, the catastrophic Irish Famine of the 1840s has been largely forgotten, save in the historical memory of the Irish. This famine was estimated to have reduced the population of Ireland by 50 percent, by causing over one to two million people to die of starvation or related diseases and another one to two million to flee Ireland in desperation. Nevertheless, it stands as a reminder of the dangers of unrestricted population growth and overdependence on a single source of food (e.g., the potato). Environmentalists ask if a similar danger exists today. A considerable part of the Earth's population relies on cereals, principally rice, maize, and wheat. Only the United States, Canada, and Australia consistently produce a surplus of these staples for export. Losses of these nation's crops to climate, mildew, rust, smut, or other blights—if great enough to prohibit the export of cereals to other countries—would lead to famines of much greater proportion than that suffered in Ireland. Even without future threats, these three countries cannot serve indefinitely as bread baskets to the world. Note, for example, that the drought-damaged United States grain harvest in 1988 fell below consumption perhaps for the first time. Furthermore, note that in 1991 the world's agricultural production actually declined for the first time since 1983. This decline occurred because of the diminishing harvest in the United States, Canada, Australia, eastern Europe, and the former Soviet Union.

Some economists, agronomists, and environmentalists, among others, predict that the world's supply of arable land, and therefore of potential food, will soon be inadequate to support future generations. The influence of technology in the form of better fertilizers, pesticides, and high-yield seed grains, as well as improved mechanization and management techniques (for food from the seas), can increase present productivity. Yet, current trends will not continue. All available evidence indicates that "as the nineties unfold, the world is facing a day of reckoning."[15] There will be a diminishing return in the use of fertilizer because many pest species are becoming resistant to pesticides. Also, high-yield plants are vulnerable to insects and diseases, and the problem of pollution by fertilizers and pesticides threatens supplies constantly. Also, the availability of water may become the single most important constraint to an increasing yield in developing countries. Moreover, most of the factors that now contribute to higher yields—fertilizer, pesticides, power for irrigation, and fuel for machinery—largely rely on fossil fuels. In addition, as exportable petroleum and other fossil fuels dwindle, food-exporting nations will be tempted to convert their exportable surpluses of grain, sugar, and other food items into biofuels.

Lester Brown points out that the growth of food production is slowing down, and he elaborates further:

If this analysis is at all close to the mark, then food scarcity is likely to emerge as the defining issue of the era now beginning, much as ideological conflict was the defining issue of the historical era that recently ended. National political leaders everywhere will be thoroughly challenged by the new demands placed on them by the prospect of growing food scarcity. Ensuring the food security of the next generation requires fundamental changes in population policy, energy policy, land use policy, water use policy, and, indeed, in the very definition of national security itself. Whether or not political leaders can respond quickly enough to avoid widespread political instability remains to be seen.[16]

Critics of the economic growth attitude also warn that the focus on growth makes ecological systems subordinate to economic systems. As the economist Herman Daly (1973) explained, "The human economy is a subsystem of the steady-state ecosystem. Therefore at some level and over some time period the subsystem must also become a steady state, at least in its physical dimensions of people and physical wealth."[17] In another essay, Daly indicates that "as the economy grows beyond its present physical scale, it may increase costs faster than benefits and initiate an era of uneconomic growth which impoverishes rather than enriches."[18]

Interestingly enough, Al Gore, Jr., who made environmental reform one of his principal concerns in the Senate and later as vice president, introduced a "new global eco-nomics" as his formula for saving the planet. In 1992 he argued:

Our challenge is to accelerate the needed change in thinking about our relationship to the environment in order to shift the pattern of our civilization to a new equilibrium—before the world's ecological system loses its current one.[19]

Kenneth Boulding, in 1969, was one of the first economists to propose the idea of a steady-state economy. To explain his rationale for a transition to a steady-state economy, he constructed a spaceship analogy of planet Earth (a closed system) to denote the finite limitations upon the resources of the Earth and the futility of consuming these resources.

Inherent in the steady-state concept is the notion that eventually the growth rate of our society will slow down and an equilibrium, or steady-state, within the environment will be more or less achieved. At this point, natural systems and human systems will be integrated. Theoretically, this last growth stage will be mature, self-maintaining, self-reproducing through its development stages, and relatively permanent. Our society will be tolerant of the environmental conditions it has imposed upon itself. This terminal society will be characterized by an equilibrium between gross industrial and agricultural production and total consumption, and between the energy captured and the energy released. There will be a balance between the consumption of natural resources and recovery of resources by treating and recycling wastes. In this final stage, more resources will be devoted to the maintenance of the complex human system than to the production of the system.

In the last stage of growth, a wide diversity and complex interaction of professionals with a well-developed spatial structure could be found. In the language of the economist, "there would be a constant stock of human capital (people) and a constant stock of physical capital (machine, building, etc.)."[20] This final stage will be the climax of our society. The human system will be similar to an ideal human system or anthroposystem, which is a self-sustainable system.

This proposed growth pattern of the human society is very similar to what occurs in natural ecosystems, where the final stage of succession is reached or approached. There are general internal changes in the ecosystem. As the ecosystem gets older, there is a tendency at any given time for a greater biomass or total quantity of living organisms. For example, trees weigh more than shrubs, shrubs weigh more than grass. A greater variety of habitats for animals are created. This leads to a greater diversity of species at the preclimax stage, as well as a tendency toward more interaction between species at the climax stage. Besides an increase in the number of terrestrial species, there is also an increase in the variety of soil organisms.

Finally, as the climax stage is reached or approached, the photosynthesis/respiration ratio approaches one. This is due to the fact that in the pioneer state, the photosynthetic rate is greater than the respiratory rate. The pioneer stage contains fairly simple food webs comprised of mostly producers. However, as the community grows older, more animals colonize the region and the ecosystem's resources are consumed mostly by respiration. Matter and energy go primarily into maintaining the existing food webs. The community is relatively more stable and self-sustainable. The successional interactions in natural ecosystems have evolved over millions of years by natural selection to create a "balance of nature."

The body growth pattern of organisms is also similar. For instance, the rate of body growth is very rapid at the beginning of an organism's life, slows down in adolescence, and becomes more stable in adulthood.

The prominent ecologist Eugene Odum was one of the first scientists to observe the parallels between development of human society and development in natural ecosystems. Among the points he made, several of which have already been described, are the following:

The shift in energy use from growth to maintenance . . . the most important trend in ecological succession has its parallel in growing cities and countries. People and governments consistently fail to anticipate that as population density increases and urban-industrial development intensifies, more and more energy, money, management effort, and tax revenues must be devoted to the services (e.g. water, sewage, transportation, and police) that maintain what is already developed and "pump out the disorder" inherent in any complex, high-energy system. Accordingly, less energy is available for new growth, which eventually can come only at the expense of the development that already exists. The transition from youth to maturity is indeed a painful and difficult time for societies, as it is for individuals, because many attitudes and goals have to be reversed.[21]

POLICIES AND PROGRAMS CONCERNING SUSTAINABLE SYSTEMS

Developing policies and programs concerning sustainable human society has become one of the most vital issues in the development of environmental jurisprudence since the 1970s. But the "law and justice" of developing compatible economic and environmental strategies remains largely underdeveloped. This is due primarily to the fact that although people can see and feel the immediate effects of pollutants, such as when some toxic substances injure people or endanger public health, the most important function of the environmental law, protecting the order of the human-environment system, tends to be overlooked. After all, without a sustainable system, there would be no society as we know it. This is where the "environmental engineering" aspect of environmental jurisprudence comes into play.

The value of protecting the order of the human-environment system, although it will not be affected for many decades to come, appears understandably insignificant to most individuals when compared with the value of protecting public health. As one commentator appropriately noted, "Global warming has yet to kill a single human being and may not do so for centuries."[22]

Worldwide, many other issues are more pressing. In developing countries, for example, the vast majority of poor people are too concerned with surviving from day to day to be worried about some vague academic notion referred to as "sustainability." Many contend that the world has been around at least since human history began, so there is no cause to believe that there will be a disorder in the foreseeable future, if ever.

It seems that protecting the order of nature and indirectly achieving sustainability is primarily a concern for those individuals who are technologically sophisticated, altruistic, future-oriented, and economically speaking, relatively better off. Consequently, the maintenance of ecological integrity is a complex function, which is more prominent in developed societies.

Many environmentalists have argued that the principal goal of protecting the order of nature can be achieved by educating the world's masses on the values of sustainability. In the United States, private environmental groups, such as the Sierra Club and the Izaak Walton League (which educate the public about environmental issues), and public agencies with environment-related duties, such as the EPA's Office of Environmental Education, have made environmental literacy one of their chief goals.

Other environmentalists warn that whatever the long-term benefits of "environmental literacy," the damages to the environment are too immediate and catastrophic to rely on education alone. A number of environmental groups have called for direct action on behalf of the environment. Among the notable organizations in this respect are Earth First! and Greenpeace. Members of these activist organizations have put themselves in harm's way by thwarting environmentally damaging practices. In Oregon, for example, Earth First! members interfered with deforestation. They stood between running bulldozers and trees, sat on company dynamite to prevent blasting, and chained themselves to timber equipment. Greenpeace members have taken their stand between whaling ships and whales to prevent the animals' slaughter.

Because scientists are not absolutely sure what actions are required to maintain the order of nature, the momentum for, and direction of, environmental policy has diffused. Scientists do not know the exact ecological or functional roles of any species. Scientific knowledge in regard to identification, measurement, and inclusion of all variables interacting and to form a stable human-environment system remains hazy. With scientists themselves unsure of how nature functions, public policymakers have remained lax in shaping policies to protect the order of nature.

During the 1990s, Western policymakers sympathetic to the idea of self-sustainable human systems—such as Norwegian Minister Gro Harlen

Brundland, Canadian diplomat Maurice Strong, and U.S. Vice President Al Gore—rallied around the guiding principle of the 1987 Report of the World Commission on Environment and Development. The report emphasizes that in order to achieve sustainability, there is a need to stimulate higher levels of economic development without inflicting irreversible damage on the environment. Thus, the sustainability viewpoint centers not just on future supplies of resources but also on the impact that the using of resources has on the environment as well as economic development and the quality of life. The necessary conditions for sustainable development as identified by the Commission require:

1. An economic system that secures effective citizen participation in decision-making.
2. An economic system that provides the solution for tensions arising from disharmonious development.
3. A production system that respects the obligation to preserve the ecological foundation for development.
4. A technological system that encourages sustainable patterns of trade and finance.
5. A global system that encourages sustainable patterns of trade and finance.
6. An administrative system that is flexible and has the ability to correct itself.

In 1992 the policy of sustainable development was further fostered by the "Earth Summit" in Rio de Janeiro. In fact, a major outcome of this summit was Agenda 21, a 580-page action plan charting principles and strategies for achieving sustainable development into the twenty-first century. Pursuant to this summit, the United Nations established a Commission on Sustainable Development to monitor Agenda 21 endeavors. According to the agenda:

In order to ensure the effective follow-up of the Conference, as well as to enhance international cooperation and rationalize the intergovernmental decision-making capacity for the integration of environment and development issues and to examine the progress in the implementation of Agenda 21 at the national, regional and international levels, a high-level Commission on Sustainable Development should be established.[23]

The Commission is a senior body of the Economic and Social Council, consisting mostly of ambassadors. As a high-ranking commission, located in New York, it promises to provide a forum for discussing global experiences in implementing the recommendations of the World Commission on Environment and Development.

During the 1990s, a number of governments worldwide outlined a national strategy based on the concept of sustainable development. In the United States, for example, the President's Council on Sustainable Development was charged in 1993 with developing new patterns to integrate environmental, economic, and social policies; reduce the cost and conflict; and garnish the benefits of environmental protection. By charter, the Council, a twenty-five-member partnership of high-level representatives from private and public organizations, was authorized to:

- Advise the president on matters involving sustainable development. In furtherance of the mission, the Council will develop and recommend to the president a national sustainable development action strategy to foster economic vitality.

- Advise the president on fashioning an annual Presidential Award recognizing exemplary efforts in advancing sustainable development ideals; submit nominations for the award to the president.

- Advise the president on conducting a public awareness and participation campaign on appropriate uses of the nation's natural and cultural resources.[24]

The preceding paragraphs reveal the caution governing environmental policy. Rather than use terms like "steady-state economy," which implied no growth, most politicians preferred the term "sustainable economic development," which suggested a self-sustainable system. The latter concept originated as early as the 1970s, when such comparable terms as "anthroposystem" and "spaceship Earth" entered the environmental lexicon (refer to Table 5.1). That which gave the sustainable movement a strong impetus was the publication in 1987 of *Our Common Future* by the United Nations World Commission on Environment and Development.

Environmental policymakers' preference for the term "sustainable development" rather than "steady-state economy" or "anthroposystem" notwithstanding, the great majority of ecologists and environmental scientists agreed in the 1990s that no technical fix alone would change the stark reality that infinite growth on a finite planet is impossible on Earth. History and circumstances, in sum, have locked human society when measured against expectations raised by human wants and needs that are ecologically unacceptable. Ecologists and environmental scientists urge society to acknowledge the constraints of the ecosphere.

Such acknowledgment, and the public policies that would evolve from it, did not occur during the twentieth century. For public decision makers to acknowledge these limits, and for society to accept the verdict, new sets of policies that would focus not only on how to make the economy grow beyond its carrying capacity but on how to progress within its ecological limits are required. If sustainable development meant progressing within ecological limits, then the anthroposystem seemed inevitable. But as several ecologists and environmentalists observed by the late 1990s, when a society reaches its carrying capacity, the distinction between self-sustainable system, anthroposystem, steady-state economy, and sustainable development, will be in degree rather than in kind.

NOTES

1. Charles H. Southwick, *Global Ecology in Human Perspective* (New York: Oxford University Press, 1996), 312.

2. Robert C. Paehlke, "Environmental Values and Policy," in Norman J. Vig and Michael E. Kraft, eds., *Environmental Policy in the 1990s* (Washington, DC: Congressional Quarterly Press, 1994), 354.

3. Donnella Meadows et al., *The Limits to Growth* (New York: Universe Books, 1972).

4. World Commission on Environment and Development, *Our Common Future* (Oxford: Oxford University Press, 1987), 1.

5. Ibid., 43.

6. Miguel A. Santos, *Ecology, Natural Resources, and Pollution* (London: Living Books, 1975), 52.

7. Council on Environmental Quality, *The Sixteenth Annual Report* (Washington, DC: Superintendent of Documents, 1985), 177.

8. Edward A. Shevardnadze, Statement Before the Forty-Third Session of the U.N. General Assembly, post verbatim 6 (September 27) (New York: United Nations, 1988), 76.

9. William R. Catton, Jr., "The World's Most Polymorphic Species," *BioScience* 37 (1987): 413–19.

10. Lester R. Brown, "The Acceleration of History," in Lester R. Brown et al., eds., *State of the World 1996* (New York: W. W. Norton & Co., 1996), 4.

11. Miguel A. Santos, "Ecological Systems versus Human Systems: Which Should be Supreme?" *Journal of Environmental Systems* 4 (1974): 261.

12. Lester R. Brown and Sandra Postel, "Thresholds of Change," in Lester R. Brown et al., eds., *State of the World 1987* (New York: W. W. Norton & Co., 1987), 4.

13. Julian L. Simon, "Bright Global Future," *The Bulletin of the Atomic Scientists* 40 (1984): 15.

14. World Resource Institute, *World Resources 1994–95* (New York: Oxford University Press, 1994), 107.

15. Lester R. Brown, "Facing Food Insecurity," in Lester R. Brown et al., eds., *State of the World 1994* (New York: W. W. Norton & Co., 1994), 177.

16. Lester R. Brown, "Facing the Prospects of Food Scarcity," in Lester R. Brown et al., eds., *State of the World 1997* (New York: W. W. Norton & Co., 1997), 41.

17. Herman E. Daly, "The Steady-State Economy: Toward a Political Economy of Biophysical Equilibrium and Moral Growth," in Herman E. Daly, ed., *Toward a Steady-State Economy* (San Francisco: Freeman & Co., 1973), 153.

18. Herman E. Daly, *Toward a New Economic Model* (Bulletin of the Atomic Scientists, April, 1986).

19. Albert Gore, Jr., *Earth in the Balance: Ecology and the Human Spirit* (Boston: Houghton Mifflin, 1992), 48.

20. R. Kerry Turner, David Pearce, and Ian Bateman, *Environmental Economics* (Baltimore: The Johns Hopkins University Press, 1993), 3.

21. Eugene P. Odum, *Ecology and Our Endangered Life-Support Systems* (Sunderland, MA: Sinauer Associates, Inc., 1989), 205.

22. Mabubul Haq, quoted in the *New Yorker*, June 29, 1992, p. 25.

23. Cited in Adam Rogers, *The Earth Summit: A Planetary Reckoning* (Los Angeles: Global View Press, 1993), 310.

24. Council on Environmental Quality, *Twenty-Fourth Annual Report* (Washington, DC: Superintendent of Documents, 1993), 331.

Biographies: The Personalities Behind the Environmental Crisis

The environmental crisis of the twentieth century is complex and pervasive. It affects all media of air, land, water, and energy, as well as our socioeconomic-political system. Resolving it requires people from many backgrounds, especially ecologists, environmental scientists, economists, and political scientists. Thousands of these concerned citizens, as individuals or as members of organizations, have been instrumental in influencing society's environmental public policy.

The objective of this chapter is to describe the key players who brought the environmental paradigm into society's forefront. We discuss their careers and how their fundamental ideologies helped focus societal attention to a specific public concern. Space limits the ability to say something about the thousands of adherents and supporters of environmental movements. Rather, we will concentrate on those environmentalists who played a seminal role in the development of an analytical framework of the environmental problem.

Kenneth E. Boulding (1910–1993)

Kenneth E. Boulding, originator of the "spaceship Earth" concept, was one of the few economists drawn into environmental politics. He was born in England, on January 18, 1910, but became a United States citizen in 1948. He graduated from Oxford University and became a professor at the University of Colorado. His research interests have included the integration of economics with political science, sociology, ecology, and other fields of study.

Boulding was one of the first economists to propose that our economy should become a steady-state economy. To explain his reasoning for a transition to a steady-state economy, he constructed an analogy of the planet Earth as a spaceship (a closed system) to denote the finite limitations upon the resources of the earth and the futility of consuming these resources. Boulding suggests:

In the spaceman economy, throughput is by no means a desiratum, and is indeed to be regarded as something to be minimized rather than maximized. The essential measure of the success of the economy is not production and consumption at all, but the nature, extent, quality, and complexity of the total capital stock, including in this the state of the human bodies and minds included in the system. In the spaceman economy, what we are primarily concerned with is stock maintenance, and any technological change which results in the maintenance of a given total stock with a lessened throughput (that is, less production and consumption) is clearly a gain.[1]

Recognized as a leading authority on the synthesis of economic theory with other disciplines, he has served as president of a number of major scholarly societies, including: American Economics Association, Society for General Systems Research, International Studies Association, American Association for the Advancement of Science, Peace Research Society International, and the Association for the Study of the Grants Economy. Boulding has written many books and articles, including his landmark essay "The Economics of Coming Spaceship Earth" (1966), in which he argued that economic growth for growth's sake is polluting and unsustainable. He died on March 19, 1993.

Lester Brown (1934–)

Lester Brown is a prominent policy-oriented natural scientist who is widely respected for his research on the global environmental problem. He was born in Bridgeton, New Jersey, on March 28, 1934. From his youth, plants interested him. During high school and college, for example, he grew tomatoes. After graduating from Rutgers University, Brown received his M.P.A. degree from Harvard University. As an expert on agricultural science and economics, he has been an outspoken supporter of human progress through sustainability. Through his long and distinguished career, Brown has served as an administrator of the International Agricultural Development Service, as policy advisor to the Secretary of Agriculture on world food needs and agricultural development abroad, as a senior fellow with the Overseas Development Council, and as president and senior researcher with the Worldwatch Institute. His publications include numerous books and articles, but he is best known for being the project director of the World-watch Institute report on progress towards a sustainable society. This annual report entitled *State of the World*, which is coauthored by outstanding contributing researchers, informs policymakers and the general public about environmental quality, and has been translated into twenty-seven languages.

Rachel Carson (1907–1964)

Rachel Carson, an American biologist, is credited for bringing public attention to the problem of environmental contamination by pesticides in the 1960s. Carson was born in Springdale, Pennsylvania, on May 27, 1907, and received her M.A. in marine biology from Johns Hopkins University in 1932. Later she taught zoology at the University of Maryland. Shortly thereafter, she became editor-in-chief of publications in the United States Fish and Wildlife Service. At the beginning of her career, her interest had focused on marine wildlife, which led her to write several books about marine science. One of her books, *The Sea Around Us* (1951), was on the United States nonfiction best-seller list for thirty-nine weeks, and won the National Book Award. In the 1950s, her work with the Fish and Wildlife Service made her aware of the growing impact of environmental pollution, especially pesticides, on biota. In 1962 *Silent Spring* alerted the public to the insidious and widespread effect. It demonstrated to the public that some pesticides were poisoning both humans and wildlife, that these toxic substances endure in the tissues long after spraying events, that they were turning up in places that had never been sprayed, and that they were endangering the delicate "balance of nature." Immediately, her book was denounced and ridiculed as being unfounded and heretical by many biologists and by the chemical industry. One reviewer said that *Silent Spring* would be avidly read by "organic gardeners, the antifluoride leaguers, the worshippers of 'natural foods,' those who cling to the philosophy of a vital principle, and pseudoscientists and faddists."[2]

Despite the criticisms, her book proved to be a best-seller. President John F. Kennedy appointed a special panel to study the problem of pesticides; the panel's report completely vindicated Carson's revelation. *Silent Spring* helped trigger the modern environmental protection measures. Unfortunately, Carson did not get to see the important historical ramification of her book, as she died of breast cancer on April 14, 1964.

Barry Commoner (1917–   )

Barry Commoner helped create the modern environmental movement by writing popular and widely read books that combined advocacy and science. Commoner was born on May 28, 1917, in New York City and earned his Ph.D. from Harvard University in 1941. Originally, his interest was focused on cellular physiology while he was a professor at Washington University in St. Louis. Concerned in the early 1950s by the health risks posed by nuclear power, Commoner's interests switched to environmental issues, helping to form a committee to publicize information about radioactive fallout from atomic testing in 1958.

He is considered one of the most vigorous champions of the faulty technology hypothesis. In his eloquent *The Closing Circle* (1971), he warned the general public of the dangerous effects of misusing technology: "Unwittingly, we have created for ourselves a new and dangerous world. We would be wise to move through it as though our lives were at stake."[3] Fundamentally, Commoner argued that technological change, with its heavy dependence on artificial energy sources and its nonbiodegradable by-products, has been the chief culprit of environmental deterioration rather than overpopulation. Consequently, in order to solve the environmental crisis, humans need to change to technologies that are environmentally benign—for example, the use of motor engines that operate at low pressure and temperature and do not generate smog-triggering nitrogen oxides, or the use of biodegradable products that naturally decompose.

Paul Ehrlich and Anne Ehrlich, both ardent advocates of zero population growth, have highly criticized Commoner's viewpoint. They remarked:

As anyone who has seen the cumulative effects of primitive "slash-and-burn" agriculture in overpopulated areas of the tropics could have informed that writer, Barry Commoner, nothing could be further from the truth.[4]

Nevertheless, Commoner has remained steadfast in his advocacy for the use of cleaner, more environmentally benign technology. Moreover, he considers Ehrlich's population regulation strategy as unsuitable, and Hardin's lifeboat ethics as uncivilized.

During the 1980 presidential elections, Commoner founded the Citizen's Party. This liberal party supported government control of the energy industry, a shift to more environmentally benign energy sources like solar energy, price control to dampen inflation, and a curb on military spending. Presently, Barry Commoner is the director of the Center for Biology of Natural Systems at Queens College (CUNY).

Paul Ehrlich (1932–   )

Paul Ehrlich was born in Philadelphia, Pennsylvania, on May 29, 1932, and grew to be an outspoken population ecologist. He helped establish the modern environmental movement by writing popular and widely read books on the issue of overpopulation. After receiving his Ph.D. from the University of Kansas in 1957, he became a professor at Stanford University. He has written numerous books and articles that reflect his lifelong interest in population biology, ecology, and evolution. However, Ehrlich is best known as an advocate of stricter control of population growth. According to him, the increase in environmental damage is proportional to the growth of population. In *Population Bomb* (1968), he alerted the public to the environmental degradation that would result if society did not better regulate the mushrooming of the human population. He prescribed population stabilization through voluntary and involuntary methods:

We must have population control at home, hopefully through a system of incentives and penalties, but by compulsion if voluntary methods fail. We must use our political power to push other countries into programs which combine agricultural development and population control.[5]

Ehrlich is one of the most dynamic and effective environmental activists of the twentieth century. In his role as an editor of population biology and consultant in biology for the McGraw-Hill Book Company, his ideas have entered many school curricula. He is also an member of the National Academy of Arts and Sciences, and a fellow of the American Academy of Arts and Sciences. In 1968 he helped to establish Zero Population Growth, an environmental group that encourages every couple to have no more than two children.

Garrett Hardin (1915–   )

Garrett Hardin is one of the twentieth century's most influential ecologists and has been at the forefront of the ethical debate over the property rights theory and the population dilemma. He was born in Dallas, Texas, on April 21, 1915, and earned a Ph.D. from Stanford University in 1941. As a professor of biology at the University of California in Santa Barbara, Hardin has made a significant impact on our understanding of the link between ethics and biology. He is best known for the classical essay, "The Tragedy of the Commons," in which he indicates that when resources, such as the global atmosphere and oceans, are owned in common, depletion of such resources is bound to occur.

The prescription offered by economists to overcome the "tragedy of the commons" is to make buyers and sellers take into account all the effects of their behaviors and so internalize the economic activity. This lets market forces place the costs of pollution on individuals who are responsible for the behavior that generated the pollution.

Hardin is also known for advocating several controversial ideas addressing the worldwide food-population concern. He has suggested that we use the notion of "lifeboat ethics" to decide with which nations we should share our resources. He points out that the world is already overpopulated and that there are too many people in the lifeboat Earth. Consequently, most nations are drowning in a sea of hopelessness. Those nations in the lifeboat should provide aid to nations that are reducing their population, not to those that are adding more people to an already overcrowded lifeboat. Sooner or later, the boat sinks, which will lead to catastrophe. In Hardin's words:

We cannot risk the safety of all the passengers by helping others in need. What happens if you share space in a lifeboat? The boat is swamped and everyone drowns. Complete justice, complete catastrophe.[6]

According to Hardin, if the developed nations want to help, they should direct their aid toward bringing population growth down. According to Hardin, "If you give food and save lives and thus increase the number of people, you increase suffering and ultimately increase the loss of life." Moreover, he strongly advocates that the United States restrict its immigration.

Aldo Leopold (1886–1948)

Aldo Leopold was holder of the first chair of game management in the United States and writer of *Game Management* (1933), the first textbook in the field. He is best known for being a proponent of "land ethics," a philosophy that holds that humans are more than managers of nature, humans are a part of nature. Leopold was born on January 11, 1886, on an Iowa farm. He earned a master's degree in forestry from Yale University in 1909. In the early years of his professional career, he was a forester for the United States Forest Service. During this period of time, he developed field experience in wilderness areas and management of wildlife forests. He began with a more traditional method by supporting the extermination of animals deemed as pests by humans, such as the timber wolf and grizzly bear. As years passed, however, Leopold began to have a deep understanding and appreciation for wildlife. He argued that we should include nature in our ethical concerns. In his words: "That land is a community is the basic concept of ecology, but that land is to be loved and respected is an extension of ethics."[7]

Later in his career, Leopold accepted a position as a professor of game management at the University of Wisconsin. By his writings and teaching, Leopold became one of the founders of the environmental movement of the twentieth century. In addition to writing his influential book *Land Ethics* (published posthumously in 1949), in which he eloquently expressed the idea that humans include nature in ethical concerns, he was also one of the founders of the Wilderness Society. Today, the Wilderness Society is a major environmental advocacy group with about one-quarter million members. He died on April 21, 1948.

James Lovelock (1919–   )

James Lovelock is best known as the originator of the Gaia hypothesis. Born in Letchworth Garden City, the United Kingdom, on July 26, 1919, he obtained a Ph.D. in 1949 from the London School of Hygiene and Topical Medicine and a D.Sci. degree in biophysics in 1959. From 1941 to 1961, Lovelock was a staff scientist at the London National Institute. In 1961 NASA recruited him to develop instruments and experiments that could be used to discover whether life exists on other planets. Lovelock compared the atmosphere of the Earth with that of Mars and noted how organisms on Earth interact with the environment in such a way that the environment might be considered an extension of the organisms. That insight led to the development of his controversial Gaia hypothesis. According to this hypothesis, the Earth is a superorganism, or in his words, "the largest living organism" at least to the extent that it can regulate itself, somehow keeping temperature, oxygen levels and other key environmental variables within narrow tolerances favorable for life. The Gaia hypothesis sees the biosphere, atmosphere, hydrosphere, and pedosphere as existing in a tightly coupled system.

Because the Gaia concept stresses that the biosphere, atmosphere, hydrosphere, and pedosphere are intimately connected by many complex pathways, it has attracted many environmentalists. However, due to its vitalistic and teleological underpinnings, the hypothesis is strongly criticized by the majority of professional ecologists. One popular ecology textbook explains the problem with the Gaia hypothesis:

It is easy to make light of the Gaia hypothesis. . . . The most serious problem, though, is the evolutionary one. The idea that the homeostatic mechanisms seen in the biosphere evolved with the specific function of maintaining a steady state is, to most evolutionary biologists, group selection pushed to an absurd extreme. What Lovelock sees as Gaia's tendency to optimize conditions for life, most biologists would see as a network of individually evolved interactions. "Gaia" is a by-product—in effect, an accident.[8]

Nevertheless, Lovelock's unorthodox ecological hypothesis inspired multidisciplinary research to validate it. He has written three books on the Gaia hypothesis: *Gaia: A New Look on Life on Earth* (1979), *The Ages of Gaia* (1988), *Healing Gaia: Practical Medicine for the Planet* (1991).

Besides proposing the Gaia hypothesis, Lovelock is a recognized independent scientist and inventor. In 1974 Lovelock was elected a Fellow of the Royal Society, and in 1990 he was awarded the first Amsterdam Prize for the Environment by the Royal Netherlands Academy of Arts and Sciences. One of his inventions is the electron capture detector, a device for use in gas chromatography that can detect pesticides and CFCs residue. Some of his inventions were adopted by NASA in their space exploration programs. The investigations to verify the existence of Gaia have occupied twenty-five years—and have been funded by the proceeds of his scientific inventions. Presently, Lovelock is retired and lives in England.

John Muir (1838–1914)

John Muir was a nature preservationist and activist who founded the Sierra Club in 1892. Muir was born on April 21, 1838, in Scotland. He and his family migrated from Scotland to a farm in Wisconsin in 1849. Muir explored much of northern America himself, keeping a journal as he traveled. He also ventured to Asia, North Africa, Australia, and New Zealand. His journeys and experiences were published in newspapers and popular and scientific magazines to a wide-ranged audience. For the most part, he lived a life that inspired his readers to the wilderness for the renewal of the wilderness mystique. In his beckoning words:

Climb the mountains and get their good tidings. Nature's peace will flow into you as sunshine flows into trees. The winds will blow their own freshness into you and the storms of energy, while cares will drop off like the autumn leaves.[9]

His philosophy was deeply spiritual, and he believed that humanity was immersed in nature rather than the master of it. During the 1890s, Muir founded the Sierra Club "to explore, enjoy, and render accessible the mountain regions of the Pacific Coast," as well as to garner support of the government in preserving these regions. Today, Muir's Sierra Club is one of the oldest, largest, and most powerful forces in environmental protection. It has approximately one-half million members, a staff of about four hundred, and an annual budget of around $20 million. Muir died on December 24, 1914.

Arne Naess (1912–   )

Arne Naess is a widely known philosopher and the founder of the deep ecology movement. Born in Norway, on January 27, 1912, Arne Naess loved climbing mountains. In 1950 he was leader of the first Norwegian Himalayan expedition. He studied in Oslo, Paris, and Vienna. After graduation, he became professor of philosophy at Oslo University. As a philosopher, mountaineer, and an outdoorsman, it was quite natural for Naess to ponder the relationship between humankind and nature. In 1972 he coined the term *deep ecology* to refer to a profound shift in the human-centered understanding of what it means to be human. The eight basic principles of the deep ecology movement are:

1. The well-being and flourishing of human and nonhuman life on Earth have value in themselves (synonyms: intrinsic value, inherent worth). These values are independent of the usefulness of the nonhuman world for human purposes.
2. Richness and diversity of life-forms contribute to the realization of these values and are also values in themselves.
3. Humans have no right to reduce this richness and diversity except to satisfy vital needs.
4. The flourishing of human life and cultures is compatible with a substantially smaller human population. The flourishing of nonhuman life requires a smaller human population.
5. Present human interference with the nonhuman world is excessive, and the situation is rapidly worsening.
6. Policies must therefore be changed. These policies affect basic economic, technological, and ideological structures. The resulting state of affairs will be deeply different from the present.
7. The ideological change will be mainly that of appreciating life quality (dwelling in situations of inherent value) rather than adhering to an increasingly higher

standard of living. There will be a profound awareness of the difference between bigness and greatness.

8. Those who subscribe to the foregoing points have an obligation directly or indirectly to try to implement the necessary changes.[10]

Deep ecologists basically argue that society faces an environmental crisis that cannot be overcome by working within established channels that divide humankind from nature. Rather, the crisis requires radical actions. According to deep ecologists, it is better to overturn society and to explore new ways of living with nature, a society in which humankind and nature are one. To deep ecologists, environmentalists who believe that human-nature relationships can be transformed within the existing social structures are dubbed "shallow ecologists." Shallow ecologists are reformists, who advocate environmental values based upon the anthropocentric ethics.

Presently, Arne Naess is professor emeritus at the University of Oslo. He is the author of many publications on empirical linguistics, Spinoza, Gandhi, and environmental philosophy.

Eugene Odum (1913–)

Eugene Odum, a highly influential policy-oriented ecologist, advocates ecologically sound public policies. Odum was born on September 17, 1913, in Lake Sunapee, New Hampshire, and earned his Ph.D. in ecology from the University of Illinois in 1939. Odum became one of the few scientific ecologists to be drawn into environmental politics. As a specialist in ecosystem ecology, he is credited for writing the most influential ecology textbook of the mid-twentieth century. *Fundamentals of Ecology* (1953) was organized around the concept of the ecosystem; stressed dynamic processes, such as biogeochemical cycles and energy flow; and integrated ecology with physics and chemistry.

In one of his most recent books, *Ecology and Our Endangered Life-Support Systems* (1989), he attempted to provide the general reader with a guide to the basic concepts of ecology as they relate to environmental issues. As an advocate of the holistic nature of ecology, he has argued that the concept of the ecosystem provides a common denominator for society and nature, and neither can be understood in isolation. He writes:

When the "study of the household" (ecology) and the "management of the household" (economics) can be merged, and when ethics can be extended to include environmental as well as human values, then we can be optimistic about the future of humankind. Accordingly, bringing together these three "E's" is the ultimate holism and the great challenge for our future.[11]

There are over 150 publications among Odum's major accomplishments, which include the following: past president of the Ecological Society of America (1964–65), Institute de la Vie Prize (1975) awarded by the French Government, Tyler Ecology Award presented by President Carter (1977), and the Crawford Prize of the Royal Swedish Academy of Sciences (1987). Presently, he is Calloway Professor Emeritus of Ecology at the University of Georgia.

Gifford Pinchot (1865–1946)

Gifford Pinchot was one of the magisterial figures in the field of conservation. As a friend of President Theodore Roosevelt and an astute politician, Pinchot became a catalyst and publicist of the anthropocentric ethics that underlie much of the twentieth century environmental policies. Pinchot was born on August 11, 1865, to an affluent family in Simsbury, Connecticut. He spent his boyhood years with his family in Connecticut and New York City. Endowed with imagination and love of nature, he shared his money, possessions, and intellect to further the cause of his utilitarian philosophy. After graduating from Yale University, he left for Europe to study forestry, because no such training existed in the United States.

Many of his anthropocentric ethics were developed during his European experience, as he began to realize that the most efficient means of exploiting the nation's forests was to develop them through careful government planning rather than private supervision. He combined his scientific skills and intellectual leadership to succeed in becoming head of the Division of Forestry in 1898. In 1905 Pinchot successfully got all the national forests transferred to his agency, by then called the U.S. Forest Service. He served as Chief Forester under Presidents William McKinley, Theodore Roosevelt, and William Taft, though he fell out of favor with Taft in disagreements on how best to set aside and manage natural resources in the federal trust. In addition to promoting conservation measures, Pinchot was governor of Pennsylvania from 1923 to 1927 and from 1931 to 1935. He died on October 4, 1946.

NOTES

1. Kenneth E. Boulding, "The Economics of the Coming Spaceship Earth" in Henry Jarret, ed., *The Economics of the Coming Spaceship Earth* (Baltimore: Johns Hopkins University Press, 1966).

2. William J. Darby, "Silence, Miss Carson," *Chemical and Engineering News* (October 1, 1962): 60.

3. Barry Commoner, *The Closing Circle* (New York: Alfred A. Knopf, 1971), 231.

4. Paul Ehrlich and Anne H. Ehrlich, *Population Resources Environment* (San Francisco: W. H. Freeman, 1970), 266.

5. Paul Ehrlich, *The Population Bomb* (New York: Ballantine, 1968), 6.

6. Garrett Hardin, "Living on a Lifeboat," *Bioscience* 24 (1974): 561–68.

7. Aldo Leopold, *A Sand County Almanac* (New York: Oxford University Press, 1949), viii.

8. Richard Brewer, *The Science of Ecology* (Troy, MO: Saunders College Publishing, 1994), 372.

9. John Muir, *Our National Parks* (Boston: Houghton Mifflin Co., 1901), 63.

10. Arne Naess, "The Deep Ecological Movement: Some Philosophical Aspects," *Philosophical Inquiry* 8 (1986): 10–31.

11. Eugene P. Odum, *Ecology and Our Endangered Life-Support Systems* (Sunderland, MA: Sinauer Associates, Inc., 1989), 271.

Primary Documents of the Environmental Crisis

PLATFORMS OF NONGOVERNMENTAL ENVIRONMENTAL ORGANIZATIONS

While the quest for environmental quality is the responsibility of the government, the unique and sizable contributions of concerned citizens working together in nonprofit environmental organizations cannot be overlooked. In fact, over the century, environmental organizations have been in the vanguard of environmental goals of various kinds. Many organizations with large memberships inform, guide, or represent their members in a wide variety of environmental and conservation efforts. For example, lawsuits brought by environmental organizations are demanding that the governmental authorities enforce environmental laws. Other organizations with similar objectives have professional memberships in recreation or conservation fields. Still other international organizations with broad interests, such as the World Wildlife Fund and Friends of the Earth, carry on many programs that have a global impact. Today, there are about one thousand environmental organizations.

To some extent, all environmental organizations are involved in public decisions about the environment. These organizations educate their constituencies and the general public as to their mission, which includes group members joining protests or disrupting the political process. Combined, the organizations are initiators of environmental policies and thereby induce policy formation to achieve environmental goals. Unlike governmental publications, environmental publications written by private organizations bring environmental issues to public attention by writing political manifestoes that may be considered too controversial for official government documents. The following is a description of the major nongovernmental environmental organizations with excerpts from their mission statements and other publications.

Document 1
DEFENDERS OF WILDLIFE

Address: 1101 14th Street, NW, #1400; Washington, DC 20005; 202-682-9400

Web site: http://www.defenders.org

Year Founded: 1959

Membership	*1969*	*1972*	*1983*	*1995*
	12,000	15,000	63,000	170,000

Major Activities: Promotes the preservation of the diversity of the world's wildlife.

Our Mission: Defenders of Wildlife is dedicated to the protection of all native wild animals and plants in their communities. We focus our programs on what scientists consider two of the most serious environmental threats to the planet: the accelerating rate of extinction of species and the associated loss of biological diversity, and habitat alteration and destruction. Long known for our leadership on endangered species issues, Defenders of Wildlife also advocates new approaches to wildlife conservation that will help keep species from becoming endangered. Our programs encourage protection of entire ecosystems and interconnected habitats while protecting predators that serve as *indicator* species for ecosystem health. (Excerpt provided by the organization.)

Document 2
ENVIRONMENTAL ACTION FOUNDATION

Address: 6930 Carroll Ave., NW; Suite 600; Takoma Park, MD 20912; 301-891-1100

Web site: http://www.social.com/health/rhic/data

Year Founded: 1970

Membership	*1972*	*1983*	*1995*
	8,000	20,000	10,000

Major Activities: Pursues strategy of political and social changes through magazines, the press, and lobbying.

Abstract: The Environmental Foundation was created in 1970 to provide local activists with the technical and organizing expertise needed to handle complex environmental issues. The Foundation provides testimony for congressional and commission hearings, leads workshops on a variety of environmental issues, trains activists on the specifics of issues, and responds by phone and mail to requests for information. Major areas of concentration are electric utilities, nuclear energy, economics, renewable energy and conservation, solid waste, hazardous waste, and toxic substances. Fact sheets available on waste reduction, recycling, resource recovery, proper disposal practices, hazardous waste management, and toxic pollution problems to local activists and state occupational health officials. (Excerpt provided by the organization.)

Document 3
ENVIRONMENTAL DEFENSE FUND

Address: 275 Park Avenue South; New York, NY 10010; Washington, DC 20009; 800-684-3322

Web site: http://www.edf.org

Year Founded: 1967

Membership	*1970*	*1980*	*1990*	*1995*
	10,000	35,000	100,000	250,000

Major Activities: The first environmental law organization dedicated to improving environmental quality and public health.

Bylaws of Environmental Defense Fund: The objectives of EDF shall include the following:

(a) to encourage and support the wise use of natural resources, and the maintenance and enhancement of environmental quality;

(b) to pursue and to encourage others to pursue research relevant to the wise use of natural resources, and the maintenance and enhancement of environmental quality;

(c) to promote interdisciplinary collaboration among scientists toward the advancement of environmental science;

(d) to effect a joining of the best scientific findings with the most appropriate social action discovered by the social sciences and legal theory in order that practical decisions shall be made which shall best promote a quality environment;

(e) to encourage public education on the problems of the wise use of natural resources and the maintenance and enhancement of environmental quality;

(f) to prevent, or to prevent the continuance of, environmental degradation by taking whatever legal measures are necessary to provide scientists fair and impartial forums in which their scientific findings may be presented objectively to their fellow citizens and tested through cross examination. (Excerpt provided by the organization.)

Document 4
FRIENDS OF THE EARTH

Address: 1025 Vermont Ave., NW; Suite 300; Washington, DC 20005-6303; 202-783-7400

Web site: http://www.foe.org

Year Founded: 1969

Membership	1972	1983	1995
	8,000	29,000	35,000

Major Activities: Committed to the preservation, the Earth restoration, and rational use of the planet.

Mission Statement: Friends of the Earth international is a worldwide federation of national environmental organizations. This federation aims to:

—protect the earth against further deterioration and restore damage inflicted upon the environment by human activities and negligence;

—preserve the earth's ecological, cultural and ethnic diversity;

—increase public participation and democratic decision making. Greater democracy is both an end in itself and is vital to the protection of the environment and the sound management of natural resources;

—achieve social, economic and political justice and equal access to resources and opportunities for men and women on the local, national, regional and international levels;

—promote environmentally sustainable development on local, national, regional and global levels.

Friends of the Earth International has a democratic structure with autonomous national groups which comply with the guidelines established by the federation.

Friends of the Earth member groups are united by a common conviction that these aims require both strong grass roots activism and effective national and international campaigning and coordination. They see Friends of the Earth international as a unique and diverse forum to pursue international initiatives, taking advantage of the variety of backgrounds and perspectives of its members.

By sharing information, knowledge, skills and resources on both a bilateral and multilateral level, Friends of the Earth groups support each other's development and strengthen their international campaigns. (Excerpt provided by the organization.)

Document 5
GREENPEACE

Address: 1436 U Street, NW; Washington, DC 20009; 202-462-1177

Web site: http://www.greenpeace.org

Year Founded: 1971

Membership	*1971*	*1980*	*1990*	*1995*
	250,000	250,000	2,500,000	1,500,000

Major Activities: Activist group concerned with such issues as nuclear weapons testing and pollution. Greenpeace members have risked their lives putting themselves in places designed to thwart environmentally damaging practices.

Greenpeace Annual Report: Greenpeace was conceived in 1971 when members of the Don't Make A Wave Committee in Vancouver, Canada, renamed their organization the better to proclaim their purpose: to create a green and peaceful world. Greenpeace today adheres to the same principle that led 12 people to sail a small boat into the US atomic test zone off Amchitka in Alaska in 1971: that determined individuals can alter the actions and purposes of even the most powerful by 'bearing witness,' that is by drawing attention to an abuse of the environment through their unwavering, presence at the scene, whatever the risk.

Policies: Greenpeace is concerned only to protect the environment. It allies itself with no political party and takes no political stance. Greenpeace is independent of the influence—financial or otherwise—of any government, group or individual. Greenpeace embraces the principle of non-violence, rejecting attacks on either people or property. (Excerpt provided by the organization.)

Document 6
IZAAK WALTON LEAGUE

Address: 707 Conservation Lane; Gaithersburg, MD 20878-2983; 800-453-5463

Web site: http://www.iwla.org

Year Founded: 1922

Membership	1960	1972	1983	1995
	51,000	56,000	47,000	53,000

Major Activities: Educates the public about emerging environmental issues, supports strong conservation laws, and through litigation, takes violators to court.

ABOUT THE IZAAK WALTON LEAGUE

Mission Statement.—To conserve, maintain, protect and restore the soil, forest, water and other natural resources of the United States and other lands; to promote means and opportunities for the education of the public with respect to such resources and their enjoyment and wholesome utilization.

Who We Are.—We are 50,000 people nationwide who care about conservation, outdoor recreation and protecting our treasured wildlife and natural resources heritage for generations to come. . . . We go about our business in a firm, polite way to oppose activities that jeopardize natural resources. As true grassroots voices throughout the United States, we are committed to promoting a centrist, commonsense approach to conservation that reflects the public's continued interest in the protection and responsible use of natural resources. All League members have a voice in the formation of League policy.

What We Are.—We engage in hands-on, local conservation action, such as monitoring streams through our Save Our Streams Program. We plant trees, restore streams, maintain trails and spend many hours teaching people about safe, ethical hunting through state hunter education courses.

Public Policy.—We educate the general public, as well as local, state and national policy-makers, about conservation and outdoor recreation issues. (Excerpt provided by the organization.)

Document 7
NATIONAL AUDUBON SOCIETY

Address: 700 Broadway; New York, NY 10003; 212-979-3000

Web site: http://www.audubon.org

Year Founded: 1905

Membership	1960	1980	1990	1995
	32,000	400,000	575,000	600,000

Major Activities: Promotes research, education, and action in protecting and managing natural resources, including wildlife species.

Strategic Plan: Though an old story, it bears re-telling—of how, a century ago, determined citizens banded together in common outrage against the slaughter of herons and egrets whose feathers were being used to adorn ladies' hats. Those citizens, who succeeded in fostering laws to protect birds and their rookeries, called themselves the Audubon Society.

The story has resonated throughout the history of the conservation and environmental movements because its guiding ethic still lies at the heart of our collective vision—an ethic which asserts that citizens can and must succeed in preserving their natural heritage.

Audubon's story parallels the story of the environmental movement, as one of its leaders, as its bellwether. In the first half of the century, Audubon was there to lead the way when protection of the nation's "crown-jewel" wildlands topped the agenda. And during the 1970s and 80s, Audubon was there at the vanguard when the nation's landmark environmental laws were enacted. But a sea of change has occurred on the environmental scene—a scene that now includes a strong opposition, a scene that reflects the nation's evolving demographics and a shift of impetus in government from Capitol Hill to the 50 state houses and, ultimately, to local communities.

Change is challenging, but Audubon's story has always been about rising to the challenge. So, to prepare ourselves to lead the environmental movement boldly forward into the new millennium, Audubon has just completed a long-term strategic plan. The process which led to the plan was thought-ful, rigorous, and disciplined—encompassing interviews, focus groups, and surveys that involved all of Audubon's staff and Directors, dozens of leading environmentalists, 700 Audubon chapter leaders, and nearly 1,300 randomly-selected Audubon members. The questions posed were tough, the answers were honest.

The resulting plan maps Audubon's course for the next 25 years, as we dedicate ourselves to bringing about a national and, yes, worldwide culture of conservation.

To accomplish this, our plan mandates that we:

• Sharpen the focus of Audubon's campaign and resources on the conser-vation of birds, other wildlife and their habitats.

• Expand our educational programs to nurture appreciation of nature and understanding of the essential link between ecological health and the well-being of human civilization.

• Invest heavily in Audubon's distinctive grassroots network as the pri-mary instrument of our environmental advocacy.

As we move forward, these initiatives will build upon our traditional strengths of education and advocacy. It will be by empowering ever-more citizens with the knowledge and means to be effective environmental advo-cates that we will bring about a culture of conservation. (Excerpt provided by the organization.)

Source: Excerpts from *Strategic Plan* used by permission of the National Audubon Society.

Document 8
NATIONAL PARK AND CONSERVATION ASSOCIATION

Address: 1776 Massachusetts Ave., NW; Washington, DC 20036; 202-223-6722

Web site: http://www.npca.org

Year Founded: 1919

Membership	1960	1980	1990	1995
	15,000	31,000	100,000	500,000

Major Activities: Protects and improves natural areas, historic cultural re-sources, and urban parks. Lobbies the government for additional funds for parks and land acquisition.

Park Advocate: NPCA identifies problems and then generates solutions, in addition to garnering the support necessary to implement those solutions. This work comprises three key activities:

Taking Action. In every region of the country, NPCA and its corps of citizen activists and allies are working to counter threats to individual parks, gaining support from interested parties for better solutions and, where necessary, taking legal action to protect natural and cultural treasures.

Advocating. On the national level, NPCA pushes for legislation that will expand, enhance, or preserve the National Park System, while working to block those bills and proposals that would dismantle the system or undermine its integrity.

Educating. NPCA educates and inspires people everywhere to appreciate the cultural and natural values of their national parks and to take a role in defending these treasures from misuse.

As the 21st century approaches, NPCA will continue its twin roles as an advocate for the national parks—supporter and watchdog. We will champion the pure, timeless values that have inspired park lovers from the beginning, while devising sophisticated new solutions and responses needed for the challenges that lie ahead. (Excerpt provided by the organization.)

Document 9
NATIONAL WILDLIFE FEDERATION

Address: 1400 Sixteenth St., NW; Washington, DC 20036; 1-800-822-9919

Web site: http://www.nwf.org/nwf/

Year Founded: 1936

Membership	*1960*	*1980*	*1990*	*1995*
	3,100,000	4,000,000	5,800,000	4,400,000

Major Activities: Area of concern is a variety of environmental topics, including conservation research and public education.

Activism, Advocacy, & Litigation: Through activism, advocacy and litigation, NWF helps people protect the wildlife, wild places and resources we all must share. Our activism begins at the grassroots level by providing our members with the knowledge and resources they need to make a difference. Through advocacy within the state and federal government, on the big and small screen, and on the front page, we bring a common-sense discussion of conservation policies to the American public. And we ensure the development and enforcement of effective conservation policies that protect the health of people, wildlife and natural resources through litigation. (Excerpt provided by the organization.)

Document 10
NATURAL RESOURCES DEFENSE COUNCIL

Address: 40 West 20th St.; New York, NY 10011; 212-727-2700

Web site: http://www.nrdc.org/

Year Founded: 1970

Membership	*1975*	*1980*	*1990*	*1995*
	18,000	29,000	138,000	170,000

Major Activities: Involved in legislative and litigation activities concerning a wide variety of environmental interests.

A Trust for the Earth: On the eve of the 21st century, humankind faces a great and overriding challenge: to pursue a new and more environmentally sustainable path, one that will heal our fragile planet for our children and our children's children. . . . There is no question that NRDC must play a leadership role in transforming humankind's relationship with the natural world. Thanks to the past support of our steadfast donors, NRDC has become widely regarded as one of the most innovative and effective environmental organizations in the world. Indeed, NRDC has helped pioneer virtually every facet of U.S. environmental law and protection over the past quarter century.

But A Trust for the Earth must reach beyond the urgent programmatic actions that our donors continue to support so generously. Where environmentalists could once tackle a leading environmental threat on a single front, we now face problems of much greater scope and complexity: worldwide and interconnected crises of ocean pollution, forest destruction, population explosion, and loss of biodiversity.

Crafting solutions to these problems will require NRDC's brand of advocacy and problem-solving on a global scale for decades to come. We must be prepared to continually advance the frontiers of energy efficiency, pollution prevention, low-input farming, recycling, and other ecologically sound technologies. We must build new partnerships with grassroots groups, farmers, fishermen, physicians, and many other constituencies for environmental solutions. And we must persevere through governments, courts, world bodies, and the private sector to establish good stewardship of the Earth as one of the central imperatives of human society. (Excerpt provided by the organization.)

Document 11
NATURE CONSERVANCY

Address: 1815 North Lynn St.; Arlington, VA 22209-2003; 703-841-8745

Web site: http://www.tnc.org

Year Founded: 1940

Membership	1970	1980	1990	1995
	114,000	182,000	630,000	553,000

Major Activities: Devoted to the protection of ecologically significant natural areas and the diversity of life they support.

Conservation by Design: A Framework for Mission Success: With the accelerating loss of the Earth's biological heritage and the diminishment of the ecosystem functions that support life on the planet, the human condition is irretrievably impoverished. The Nature Conservancy's mission therefore could not be more important or compelling.

This Framework translates our broadly stated mission into a more definitive and specific articulation of common purpose and direction: a compass setting to align and unify the whole organization in taking the most effective conservation action. It sets forth an explicit set of guidelines that enable us to design what mission success should look like everywhere and at every scale that we work. It also enables us to develop and implement strategies that create the greatest likelihood of realizing that vision. We call this approach "conservation by design."

Through this approach, we harness the full vigor of the innovative and enterprising spirit that is the hallmark of the Conservancy. With each of our state and country programs acting on a shared understanding of what constitutes success, we take full advantage of our decentralized organizational structure. Such a design also establishes benchmarks against which we can measure progress and performance. It is to this Framework, therefore, that we hold ourselves individually and collectively accountable.

We are emboldened in fulfilling this commitment by our past accomplishments, and by the understanding that wherever we work in the world we share a set of core values and distinguishing qualities that enable us to be singularly effective:

We are guided by the best available conservation science to take site-based action that makes a significant and lasting difference.

We work in a non-confrontational manner, emphasizing the effectiveness of collaborative efforts.

We recognize the imperative of developing ways to enable humans to live productively and sustainably while conserving biological diversity.

We move forward eagerly and with confidence, inspired by the boldness and importance of our mission; propelled by our past successes; distinguished by our values and unique organizational style; and guided by the direction set by this Framework. For in the end, our society will be defined not only by what we create but by what we refuse to destroy. (Excerpt provided by the organization.)

Source: Excerpts from *Conservation by Design: A Framework for Mission Success* used by permission of the Nature Conservancy.

Document 12
SIERRA CLUB

Address: 85 2nd St., Second Floor; San Francisco, CA 94105; 415-977-5653

Web site: http://www.sierraclub.org

Year Founded: 1892

Membership	*1960*	*1980*	*1990*	*1995*
	15,000	182,000	630,000	550,000

Major Activities: Lobbies and educates on many environmental issues.

Take One Step Towards A Better Planet: It began more than a century ago in the rugged wilderness of the Sierra Nevada. Deep among the towering sequoias and cascading waterfalls, John Muir and a group of conservationists laid the foundation for an organization dedicated to preserving and protecting the magnificent wildlands of this spectacular mountain range. The impact of the newly formed Sierra Club was immediate. And its efforts, long overdue.

Today, the Sierra Club numbers more than half a million strong and is the leader in the grassroots environmental movement. Sierra Club members are your neighbors and friends. They're people active in your local communities. In the past one hundred years, the goals of the Sierra Club have expanded to include not only preserving what remains of our unspoiled wilderness, but also protecting Earth's endangered species, cleaning the air we breathe and the water we drink, and ensuring a safer, healthier planet for future generations to enjoy. . . . Our goal is environmental protection but not just for trees in the forest or the fish in the streams. Ultimately, the Sierra Club is about protecting people. (Excerpt provided by the organization.)

Document 13
WILDERNESS SOCIETY

Address: 900 Seventeenth St., NW; Washington, DC 20006-2596; 202-833-2300

Web site: http://www.wilderness.org

Year Founded: 1935

Membership	1970	1980	1990	1995
	44,000	35,000	350,000	275,000

Major Activities: Educates the public on various aspects of public land system management and economic issues related to the public lands.

Mission: The Wilderness Society is a nonprofit conservation group dedicated to the protection of America's wild lands and wildlife. As part of that mission, The Society works to foster a land ethic. We rely on a combination of public education, advocacy, and economic and ecological analysis. The Wilderness Society recognizes the importance of broad ecosystem management and biodiversity and is building alliances with new constituencies. (Excerpt provided by the organization.)

Document 14
WORLDWATCH INSTITUTE

Address: 1776 Massachusetts Ave., NW; Washington, DC 20036-1904; 202-452-1999

Web site: http://www.worldwatch.org

Year Founded: 1974

Staff: 32 (nonmembership research society)

Major Activities: Informs policymakers and the general public about the damage done by the world economy to its environmental support system.

Mission: Our reason for being is to foster the evolution of an environmentally sustainable society, one in which human needs are met in ways that do not threaten the health of the natural environment or the prospects of future generations.

We seek to achieve this goal through conducting interdisciplinary non-partisan research on emerging global environmental issues, the results of which are disseminated throughout the world. In a sentence, our mission is to raise public awareness of global environmental threats to the point where it will support effective policy responses. The beneficiaries of our work range from individuals, who use our research to guide their political actions or make life-style choices, to large organizations, such as national governments, large and small businesses, international development organizations, U.N. agencies, and nongovernmental groups. (Excerpt provided by the organization.)

Document 15
WORLD WILDLIFE FUND

Address: 1250 24th St., NW; Washington, DC 20037; 202-293-4800

Web site: http://www.worldwildlife.org

Year Founded: 1935

Membership	*1985*	*1991*	*1995*
	172,000	1,000,000	5,000,000

Major Activities: Promotes worldwide efforts to protect endangered wildlife.

A History of WWF: In just over three decades, WWF-World Wide Fund For Nature (formerly known as the World Wildlife Fund) has become the world's largest and most respected independent conservation organization. With almost five million supporters distributed throughout five continents,

24 National Organizations (NOs), 5 Associates, and 26 Programme Offices, WWF can safely claim to have played a major role in the evolution of the international conservation movement.

Since 1985, WWF has invested over US $1,165 million in more than 11,000 projects in 130 countries. All these play a part in the campaign to stop the accelerating degradation of Earth's natural environment, and to help its human inhabitants live in greater harmony with nature. (Excerpt provided by the organization.)

Document 16
ZERO POPULATION GROWTH

Address: 1400 Sixteenth St., NW; Washington, DC 20036; 202-332-2200

Web site: http://www.zpg.org

Year Founded: 1968

Membership	1970	1980	1990	1995
	3,000	11,900	22,550	58,000

Major Activities: Works to achieve sustainable balance between population, resources and the environment.

ZPG FACT SHEET
Mission Statement

Zero Population Growth is a national nonprofit organization working to slow population growth and achieve a sustainable balance between the Earth's people and its resources. We seek to protect the environment and ensure a high quality of life for present and future generations. ZPG's education and advocacy programs aim to influence public policies, attitudes, and behavior on national and global population issues and related concerns.

Approach

ZPG recognizes that broad social, economic and political changes may be necessary to slow population growth. We endorse and actively support methods which are voluntary and which are positive enhancements of human rights and conditions.

ZPG condemns any use of force or violence. ZPG condemns racism in all of its forms. ZPG will not support or tolerate being knowingly associated with organizations which support or promote the use of force or violence or which espouse racism or racist beliefs.

Specific Population Issues

The United States should assume a leadership role in international efforts to slow population growth and should set an example by adopting a national population policy which commits the United States to this goal. . . . ZPG believes that quality family planning services should be made available to all people who desire such services. . . . ZPG therefore supports laws and social practices that ensure access for all women to medically safe and affordable abortion services. . . . It is ZPG's view that immigration pressures on the U.S. population are best relieved by addressing factors which compel people to leave their homes and families and emigrate to the United States. Foremost among these are population growth, economic stagnation, environmental degradation, poverty, and political repression. ZPG believes unless these problems are successfully addressed in the developing nations of the world, no forcible exclusion policy will successfully prevent people from seeking to relocate into the United States. ZPG, therefore, calls on the United States to focus its foreign aid on population, environmental, social, education, and sustainable development programs. . . . ZPG supports the removal of all incentives and subsidies for procreation and larger families. (Excerpt provided by the organization.)

Source: Excerpts from *ZPG Fact Sheet* used by permission of Zero Population Growth.

ENVIRONMENTAL STATUTES

Federal actions to control pollution are of recent origin. Prior to the 1970s, most cases of pollution were dealt with under the common law tort theories. Beginning in the 1950s, concern for environmental quality increased, and many state and local governments enacted laws in an effort to protect human health and the environment. During this period, the role of the national government was limited primarily to directing the states' attention to the pollution problem and also providing them with economic and technical assistance. Since the 1960s, the national government has either supplemented state regulations or has taken primary responsibilities to control the quality of the nation's environment.

Congressional efforts to protect human health and the environment were primarily due to the political activities of environmentalists, ecologists, and the public. As a result of these actions taken by these concerned citizens, Congress has enacted numerous laws. Some of the notable legislation includes: Federal Insecticide, Fungicide, and

Rodenticide Control Act (1947, amended and revised in 1978 and 1988); Multiple Use Sustained Yield Act (1960); Wilderness Act (1964); Solid Waste Disposal Act (1965); Clean Air Acts (1965, 1970, 1977); Species Conservation Act (1966); Wild and Scenic River Act (1968); National Environmental Policy Act (1969); Ocean Dumping Act (1972); Federal Water Pollution Control Act (1972); Noise Control Act (1972); Marine Protection, Research, and Sanctuaries Act (1972); National Coastal Zone Management Acts (1972, 1980); Endangered Species Act (1973); Safe Drinking Water Act (1974, 1986); Forest Reserve Management Acts (1974, 1976); National Forest Management Act (1976); Toxic Substances Control Act (1976); Resource Conservation and Recovery Act (1976, 1984); Surface Mining Control and Reclamation Act (1977); Resource Recovery Act (1977); Clean Water Act (1977); Endangered American Wilderness Act (1978); Quiet Communities Act (1978); National Energy Acts (1978, 1980); Comprehensive Environmental Response, Compensation, and Liability Act (1980, 1986); Alaskan Land-Use Bill (1980); and Oil Pollution Act (1990). Table 1 lists the major federal agencies responsible for administering these laws.

Table 1
Federal Agencies with Environment-Related Duties

Government Unit or Agency (year established)	Duties
Council on Environmental Quality (1970)	Advice to the president and Congress on environmental matters
Department of Agriculture (1862)	Forestry, soil conservation, research, and wilderness areas
Department of Commerce (1913)	Oceanic and atmospheric monitoring and research, management of living marine resources, coastal zone and marine sanctuary management, marine pollution, response and damage assessment; technology and export promotion
Department of Defense (1949)	Civil works construction, dredge and fill permits, pollution control and resource management at Defense facilities
Department of Energy (1977)	Energy policy coordination, research, and development
Department of Health and Human Services (1953)	Public and environmental health-related concerns

Table 1 (continued)

Government Unit or Agency (year established)	Duties
Department of Housing and Urban Development (1965)	Housing, urban parks, urban planning
Department of Interior (1849)	National parks, national wildlife refuges, Bureau of Land Management, public lands, American Indian trust lands, water resources, fish and wildlife, Outer Continental Shelf, Exclusive Economic Zone, energy and minerals, mined land reclamation, mine health and safety research, natural hazards, and geological survey
Department of Justice (1870)	Environmental law and litigation
Department of Veterans Affairs (1930)	Hospitals and veterans' health
Environmental Protection Agency (1970)	Air and water pollution, solid and hazardous waste, radiation, pesticides, noise, toxic substances, health, and environmental education
Food and Drug Administration (1906)	Administration of laws to prohibit distribution of adulterated, misbranded, or unsafe food and drugs
National Aeronautics and Space Administration (1958)	Atmospheric and space science research and coordination
National Science Foundation (1950)	Science research and coordination
Nuclear Regulatory Commission (1974)	Nuclear power licensing and regulation
Occupational Safety and Health Administration (1970)	Inspection and standards enforcement to ensure all workers a safe and healthy work environment
Smithsonian Institution (1846)	Environmental education, historical preservation, and research
Tennessee Valley Authority (1933)	Electric power generation and water resource management

Document 17
THE CLEAN WATER ACT, 33 U.S.C. §121 ET SEQ.

The principal national law regulating water pollution is the Federal Water Pollution Control Act, originally signed in 1952 and amended in 1972. The act was amended in 1977 and renamed the Clean Water Act. As amended again in 1981 and 1987, the Clean Water Act has highly ambitious goals. It provides a comprehensive framework of pollution control standards and technical and economic assistance to deal with the many stressors that affect water quality.

EFFLUENT LIMITATIONS

(a) Illegality of pollutant discharges except in compliance with law

Except as in compliance with this . . . the discharge of any pollutant by any person shall be unlawful.

(b) Timetable for achievement of objectives

In order to carry out the objective of this chapter there shall be achieved—

(1) (A) not later than July 1, 1977, effluent limitations for point sources, other than publicly owned treatment works, (i) which shall require the application of the best practicable control technology currently available as defined by the Administrator [the EPA requires point sources of pollutants to install pollution control devices that are the most practical for the firm under the circumstances] or (ii) in the case of a discharge into a publicly owned treatment works . . . ; and

(B) for publicly owned treatment works . . . effluent limitations based upon secondary treatment . . . ; or,

(C) not later than July 1, 1977, any more stringent limitation, including those necessary to meet water quality standards, treatment standards, or schedules of compliance, established pursuant to any State law or regulations . . . or required to implement any applicable water quality standard. . . .

(2) (A) . . . [for toxic and nonconventional pollutants], effluent limitations for categories and classes of point sources, other than publicly owned treatment works, which (i) shall require application of the best available technology economically achievable [polluters must install the most effective water-pollution device independent of cost] . . . which will result in reasonable further progress toward the national goal of eliminating the discharge of all pollutants. . . . Such effluent limitations shall require the elimination of discharges of all pollutants if the Administrator finds . . . that such elimination is technologically and economically achievable. . . .

(E) for conventional pollutants [such as suspended solids and fecal coliform], as expeditiously as practicable but in no case later than . . . March 31, 1989, compliance with effluent limitations for categories and classes of point sources, other than publicly owned treatment works, which . . . shall require application of the best conventional pollutant control technology. . . .

(c) Modification of timetable

(1) The Administrator may modify the requirements . . . of this section with respect to any point source for which a permit application is filed . . . upon a showing by the owner or operator of such point source . . . that such modified requirements (1) will represent the maximum use of technology within the economic capability of the owner or operator; and (2) will result in reasonable further progress toward the elimination of the discharge of pollutants. . . .

(f) Illegality of discharge of radiological, chemical, or biological warfare agents, high-level radioactive waste, or medical waste.

Notwithstanding any other provisions of this chapter it shall be unlawful to discharge any radiological, chemical, or biological warfare agent, any high-level radioactive waste, or any medical wastes into the navigable waters.

(g) Modifications for certain nonconventional pollutants

(1) General authority

The Administrator, with the concurrence of the State, may modify the requirements . . . with respect to the discharge from any point source of . . . [certain nonconventional pollutants].

(2) Requirements for granting modification

A modification under this subsection shall be granted only upon a showing by the owner or operator of a point source . . . that—

(A) such modified requirements will [use best practicable control technology or meet water quality standards]; . . .

(c) such modification will not interfere with the attainment or maintenance of that water quality which shall assure protection and propagation of a balanced population of shellfish, fish, and wildlife, and allow recreational activities, in and on the water and such modification will not result in the discharge of pollutants in quantities which may reasonably be anticipated to pose an unacceptable risk to human health or the environment because of bioaccumulation, persistency in the environment, acute toxicity, chronic toxicity (including carcinogenicity, mutagenicity or teratogenicity), or synergistic propensities.

CWA 302 WATER QUALITY RELATED EFFLUENT LIMITATIONS

(b) Modifications of effluent limitations

(2) Permits

(A) No reasonable relationship

The Administrator, with the concurrence of the State, may issue a permit which modifies the effluent limitations . . . for pollutants other than toxic pollutants if applicant demonstrates at such hearing that (whether or not technology or other alternative control strategies are available) there is no reasonable relationship between the economic and social costs and the benefits to be obtained (including attainment of the objective of this chapter) from achieving such limitation.

CWA 303 WATER QUALITY STANDARDS AND IMPLEMENTATION PLANS

(a) Existing water quality standards

(2) Any State which . . . has adopted, pursuant to its own law, water quality standards applicable to intrastate waters shall submit such standards to the Administrator. . . .

(d) Identification of areas with insufficient controls; maximum daily load; certain effluent limitations revision

(1) (A) Each State shall identify those waters within its boundaries for which the effluent limitations . . . are not stringent enough to implement any water quality standard applicable to such waters. The State shall establish a priority ranking for such waters, taking into account the severity of the pollution and the uses to be made of such waters. . . .

CWA 304 INFORMATION AND GUIDELINES

(a) Criteria development and publication

(1) The Administrator . . . shall develop and publish . . . (and from time to time thereafter revise) criteria for water quality accurately reflecting the latest scientific knowledge

(b) Effluent limitation guidelines

For the purpose of adopting or revising effluent limitations under this chapter the Administrator shall . . . publish . . . regulations, providing guidelines for effluent limitations, and, at least annually thereafter, revise, if appropriate such regulations. . . .

(d) Protection from more stringent standards

Notwithstanding any other provision of this chapter, any point source . . . which is . . . constructed . . . to meet all applicable standards of performance shall not be subject to any more stringent standard of performance during a ten-year period on the date of completion of such construction or during the period of depreciation or amortization of such facility . . . whichever period ends first. . . .

CWA 402 NATIONAL POLLUTANT DISCHARGE ELIMINATION SYSTEM

(a) Permits for discharge of pollutants
. . . the Administrator may, after opportunity for public hearing, issue a permit for the discharge of any pollutant, or combination of pollutants . . . upon condition that such discharge will meet [national effluent-limitation standards]. . . .

CWA 502 DEFINITIONS

(6) The term "pollutant" means dredged spoil, solid waste, incinerator residue, sewage, garbage, sewage sludge, munitions, chemical waste, biological materials, radioactive materials, heat, wrecked or discarded equipment, rock, sand, cellar dirt and industrial, municipal, and agricultural waste discharged into water.

Document 18
THE CLEAN AIR ACT, 42 U.S.C. 7401–7626

The Clean Air Act was signed into law in 1955, but it was the enactment of amendments to the law in 1963, 1970, 1977, and 1990 that proved turning points in progress on air pollution control. The basic objective of the Clean Air Act is to protect people and natural resources from airborne pollutants that could endanger public health and welfare.

TITLE I—AIR POLLUTION PREVENTION AND CONTROL

Part A—Air Quality and Emission Limitations

Findings and Purposes
Sec. 101. (a) The Congress finds—
(1) that the predominant part of the Nation's population is located in its rapidly expanding metropolitan and other urban areas, which generally cross the boundary lines of local jurisdictions and often extend into two or more States;
(2) that the growth in the amount and complexity of air pollution brought about by urbanization, industrial development, and the increasing use of motor vehicles, has resulted in mounting dangers to the public health and

welfare, including injury to agricultural crops and livestock, damage to and the deterioration of property, and hazards to air and ground transportation;

(3) that air pollution prevention (that is, the reduction or elimination, through any measures, of the amount of pollutants produced or created at the source) and air pollution control at its source is the primary responsibility of States and local governments; and

(4) that Federal financial assistance and leadership is essential for the development of cooperative Federal, State, regional, and local programs to prevent and control air pollution.

(b) The purposes of this title are—

(1) to protect and enhance the quality of the Nation's air resources so as to promote the public health and welfare and the productive capacity of its population;

(2) to initiate and accelerate a national research and development program to achieve the prevention and control of air pollution;

(3) to provide technical and financial assistance to State and local governments in connection with the development and execution of their air pollution prevention and control programs; and

(4) to encourage and assist the development and operation of regional air pollution prevention and control programs.

(c) POLLUTION PREVENTION.—A primary goal of this Act is to encourage or otherwise promote reasonable Federal, State, and local governmental actions, consistent with the provisions of this Act, for pollution prevention. . . .

Air Quality Control Regions

Sec. 107. (a) Each State shall have the primary responsibility for assuring air quality within the entire geographic area comprising such State by submitting an implementation plan for such State which will specify the manner in which primary and secondary ambient air quality standards will be achieved and maintained within each air quality control region in such State. . . .

Air Quality Criteria and Control Techniques

Sec. 108. (a)(1) For the purpose of establishing national primary and secondary ambient air quality standards, the Administrator shall within 30 days after the date of enactment of the Clean Air Amendments of 1970 publish, and shall from time to time thereafter revise, a list which includes each air pollutant— . . .

National Ambient Air Quality Standards

Sec. 109. (a)(1) The Administrator—

(A) within 30 days after the date of enactment of the Clean Air Amendments of 1970, shall publish proposed regulations prescribing a national primary ambient air quality standard and a national secondary ambient air quality standard for each air pollutant for which air quality criteria have been issued prior to such date of enactment; . . .

Implementation Plans

Sec. 110. (A)(1) Each State shall, after reasonable notice and public hearings, adopt and submit to the Administrator, within 3 years (or such shorter period as the Administrator may prescribe) after the promulgation of a national primary ambient air quality standard (or any revision thereof) under section 109 for any air pollutant, a plan which provides for implementation, maintenance, and enforcement of such primary standard in each air quality control region (or portion thereof) within such State. . . .

(2) Each implementation plan submitted by a State under this Act shall be adopted by the State after reasonable notice and public hearing. . . .

Sec. 112. HAZARDOUS AIR POLLUTANTS.

(a) DEFINITIONS.—For purposes of this section, except subsection (r)—

(1) MAJOR SOURCE.—The term "major source" means any stationary source or group of stationary sources located within a contiguous area and under common control that emits or has the potential to emit considering controls, in the aggregate, 10 tons per year or more of any hazardous air pollutants or more of any combination of hazardous air pollutants. . . .

(7) ADVERSE ENVIRONMENTAL EFFECT.—The term "adverse environmental effect" means any significant and widespread adverse effect, which may reasonably be anticipated, to wildlife, aquatic life, or other natural resources, including adverse impacts on populations of endangered or threatened species or significant degradation of environmental quality over broad areas. . . .

(d) EMISSION STANDARDS.—

(1) IN GENERAL.—The administrator shall promulgate regulations establishing emission standards for each category or subcategory of major sources and area sources of hazardous air pollutants listed. . . .

(2) STANDARDS AND METHODS.—Emissions standards promulgated under this subsection and applicable to new or existing sources of hazardous air pollutants shall require the maximum degree of reduction in emissions of the hazardous air pollutants subject to this section. . . .

(3) NEW AND EXISTING SOURCES.—The maximum degree of reduction in emissions that is deemed achievable for new sources in a cate-

gory or subcategory shall not be less stringent than the emission control that is achieved in practice by the best controlled similar source, as determined by the Administrator. . . .

(4) HEALTH THRESHOLD.—With respect to pollutants for which a health threshold has been established, the Administrator may consider such threshold level, with an ample margin of safety, when establishing emission standards under this subsection. . . .

(6) REVIEW AND REVISION.—The Administrator shall review, and revise as necessary (taking into account developments in practices, processes, and control technologies), emission standards promulgated under this section no less often than every 8 years. . . .

International Air Pollution

Sec. 115. (a) Whenever the Administrator, upon receipt of reports, surveys or studies from any duly constituted international agency has reason to believe that any air pollutant or pollutants emitted in the United States cause or contribute to air pollution which may reasonably be anticipated to endanger public health or welfare in a foreign country or whenever the Secretary of State requests him to do so with respect to such pollution which the Secretary of State alleges is of such a nature, the Administrator shall give formal notification thereof to the governor of the State in which such emissions originated. . . .

Part C—Prevention of Significant Deterioration of Air Quality

Subpart 1: Purposes

Sec. 160. The purposes of this part are as follows:

(1) to protect public health and welfare from any actual or potential adverse effect which in the Administrator's judgment may reasonably be anticipated to occur from air pollution or from exposures to pollutants in other media, which pollutants originate as emissions to the ambient air, notwithstanding attainment and maintenance of all national ambient air quality standards;

(2) to preserve, protect, and enhance the air quality in national parks, national wilderness areas, national monuments, national seashores, and other areas of special national or regional natural, recreational, scenic, or historic value;

(3) to insure that economic growth will occur in a manner consistent with the preservation of existing clean air resources;

(4) to assure that emissions from any source in any State will not interfere with any portion of the applicable implementation plan to prevent significant deterioration of air quality for any other State; and

(5) to assure that any decision to permit increased air pollution in any area to which this section applies is made only after careful evaluation of all the consequences of such a decision and after adequate procedural opportunities for informed public participation in the decisionmaking process. . . .

TITLE II—EMISSION STANDARDS FOR MOVING SOURCES

Part A—Motor Vehicle Emission and Fuel Standards

Establishment of Standards

Sec. 202. (a) Except as otherwise provided in subsection (b)—

(1) The Administrator shall by regulation prescribe (and from time to time revise) in accordance with the provisions of this section, standards applicable to the emission of any air pollutant from any class or classes of new motor vehicles or new motor vehicle engines, which in his judgement cause, or contribute to, air pollution which may reasonably be anticipated to endanger public health or welfare. . . .

Citizen Suits

Sec. 304. (a) Except as provided in subsection (b), any person may commence a civil action on his own behalf—

(1) against any person (including (i) the United States, and (ii) any other governmental instrumentality or agency to the extent permitted by the Eleventh Amendment to the Constitution). . . .

(2) against the Administrator where there is alleged a failure of the Administrator to perform any act or duty under this Act which is not discretionary with the Administrator, or

(3) against any person who proposes to construct or constructs any new or modified major emitting facility without a permit. . . .

Document 19
THE RESOURCE CONSERVATION AND RECOVERY ACT, 42 U.S.C. 6901–6991i

The historical roots of the Resource Conservation and Recovery Act (RCRA) began in 1965 with enactment of the Solid Waste Disposal Act, which was essentially intended to assist state and local governments in improving their capabilities of disposing of municipal solid wastes. This statute was then reauthorized in 1970 with a change in emphasis on such wastes. During this 1970 reauthorization, a report on the problems posed by hazardous waste was first required. Due to these problems, it had become clear that the greatest concern was, in fact, hazardous wastes. Congress responded by passing the RCRA of 1976 to protect human health and the environment, to reduce waste, and to conserve energy. The RCRA was subsequently amended by the Hazardous and Solid Waste Amendment of 1984.

TITLE II: SOLID WASTE DISPOSAL

Subtitle A—General Provisions

Congressional Findings

Sec. 1002. (a) SOLID WASTE.—The Congress finds with respect to solid waste—

(1) that the continuing technological progress and improvement in methods of manufacture, packaging, and marketing of consumer products has resulted in an ever-mounting increase, and in a change in the characteristics, of the mass material discarded by the purchaser of such products;

(2) that the economic and population growth of our Nation, and the improvements in the standard of living enjoyed by our population, have required increased industrial production to meet our needs, and have made necessary the demolition of old buildings, the construction of new buildings, and the provision of highways and other avenues of transportation, which, together with related industrial, commercial, and agricultural operations, have resulted in a rising tide of scrap, discarded, and waste materials;

(3) that the continuing concentration of our population in expanding metropolitan and other urban areas has presented these communities with serious financial, management, intergovernmental, and technical problems in the disposal of solid wastes resulting from the industrial, commercial, domestic, and other activities carried on in such areas;

(4) that while the collection and disposal of solid wastes should continue to be primarily the function of State, regional, and local agencies, the problems of waste disposal as set forth above have become a matter national in scope and in concern and necessitate Federal action through financial and technical assistance and leadership in the development, demonstration, and application of new and improved methods and processes to reduce the amount of waste and unsalvageable materials and to provide for proper and economical solid waste disposal practices.

(b) ENVIRONMENT AND HEALTH.—The Congress finds with respect to the environment and health, that—

(1) although land is too valuable a national resource to be needlessly polluted by discarded materials, most solid waste is disposed of on land in open dumps and sanitary landfills;

(2) disposal of solid waste and hazardous waste in or on the land without careful planning and management can present a danger to human health and the environment;

(3) as a result of the Clean Air Act, the Water Pollution Control Act, and other Federal and State laws respecting public health and the environment, greater amounts of solid waste (in the form of sludge and other pollution treatment residues) have been created. Similarly, inadequate and environmentally unsound practices for the disposal or use of solid waste have created greater amounts of air and water pollution and other problems for the environment and for health; . . .

Objectives and National Policy

Sec. 1003. (a) OBJECTIVES.—The objectives of this Act are to promote the protection of health and environment and to conserve valuable material and energy resources. . . .

Subtitle C—Hazardous Waste Management

Identification and Listing of Hazardous Waste

Sec. 3001. (a) CRITERIA FOR IDENTIFICATION OR LISTING.—Not later than eighteen months after the date of the enactment of this Act, the Administrator shall, after notice and opportunity for public hearing, and after consultation with appropriate Federal and State agencies, develop and promulgate criteria for identifying the characteristics of hazardous waste, and for listing hazardous waste, which should be subject to the provisions of this subtitle, taking into account toxicity, persistence, and degradability in nature, potential for accumulation in tissue, and other related factors such as flammability, corrosiveness, and other hazardous characteristics. . . .

Federal Enforcement

Sec. 3008. (a) COMPLIANCE ORDERS.—(1) Except as provided in paragraph (2), whenever on the basis of any information the Administrator determines that any person has violated or is in violation of any requirement of this subtitle, the Administrator may issue an order assessing a civil penalty for any past or current violation, requiring compliance immediately or within a specified time period, or both, or the Administrator may commence a civil action in the United States district court in the district in which the violation occurred for appropriate relief, including a temporary or permanent injunction. . . . (a) CRIMINAL PENALTIES—Any person who—(1) knowingly transports or causes to be transported any hazardous waste identified or listed under this subtitle to a facility which does not have a permit . . . shall, upon conviction, be subject to a fine of not more than $50,000 for each day of violation, or imprisonment not to exceed two years.

Citizen Suits

Sec. 7002. (a) IN GENERAL.—Except as provided in subsection (b) or (c) of this section, any person may commence a civil action on his own behalf.

Document 20
THE NATIONAL ENVIRONMENTAL POLICY ACT, 42
U.S.C. S/S 4321 ET SEQ. (1969)

One of the first laws written in response to the call for a national environmental policy came in the form of the National Environmental Policy Act (NEPA) of 1969. With the enactment of NEPA, the United States began a Magna Carta–like federal program that managed environmental protection. Before the 1970s, many different types of specific environmental legislation had been passed (e.g., the Rivers and Harbors Appropriation Act of 1899, the Public Health Service Acts of 1912, the Water Pollution Control Act of 1948), but earlier environmental legislation tended to be piecemeal responses to individual concerns. During the 1950s and 1960s, ecologists, environmental scientists, other scientists, public decision makers, and the public began to see many facets of the environment as a nationwide concern. The growing interest in the environment led to the enactment of NEPA, and a significant reevaluation of any other existing environmental statutes.

Before the development of NEPA, the federal role in environmental regulation was limited to the management of publicly owned lands and specific environmental statutes. By the end of the 1970s, the national government enacted and enforced regulations in many areas including occupational health and safety, resource recovery, water quality, air quality, pesticides, toxic chemicals, hazardous wastes, mine safety, coastal zones, ocean pollution, the outer continental shelf, and the upper atmosphere.

Perhaps more than any other statute, NEPA has always been and still continues to be an important and effective means of integrating environmental impacts into the fabric of federal decision making. The Act has two main principles. The first one is to ensure that federal decision makers consider the environmental impact of their actions. The second one is to provide a means by which the public is informed of and can participate in the analysis of proposed actions that affect the environment.

PURPOSE

Sec. 2. The purposes of this Act are: To declare a national policy which will encourage productive and enjoyable harmony between man and his environment; to promote efforts which will prevent or eliminate damage to the environment and biosphere and stimulate the health and welfare of man; to enrich the understanding of the ecological systems and natural resources important to the Nation; and to establish a Council on Environmental Quality.

TITLE I: DECLARATION OF NATIONAL ENVIRONMENTAL POLICY

Sec. 101. (a) The Congress, recognizing the profound impact of man's activity on the interrelations of all components of the natural environment, particularly the profound influences of population growth, high-density urbanization, industrial expansion, resource exploitation, and new and expanding technological advances and recognizing further the critical importance of restoring and maintaining environmental quality to the overall welfare and development of man, declares that it is the continuing policy of the Federal Government, in cooperation with State and local governments, and other concerned public and private organizations, to use all practicable means and measures, including financial and technical assistance, in a manner calculated to foster and promote the general welfare, to create and maintain conditions under which man and nature can exist in productive harmony, and fulfill the social, economic, and other requirements of present and future generations of Americans.

(b) In order to carry out the policy set forth in this Act, it is the continuing responsibility of the Federal Government to use all practicable means, consistent with other essential considerations of national policy, to improve and coordinate Federal plans, functions, programs, and resources to the end that the Nation may—

(1) fulfill the responsibilities of each generation as trustee of the environment for succeeding generations;

(2) assure for all Americans safe, healthful, productive, and aesthetically and culturally pleasing surroundings;

(3) attain the widest range of beneficial uses of the environment without degradation, risk to health or safety, or other undesirable and unintended consequences;

(4) preserve important historic, cultural and natural aspects of our national heritage, and maintain, wherever possible, an environment which supports diversity, and variety of individual choice;

(5) achieve a balance between population and resource use which will permit high standards of living and a wide sharing of life's amenities; and

(6) enhance the quality of renewable resources and approach the maximum attainable recycling of depletable resources.

(c) The Congress recognizes that each person should enjoy a healthful environment and that each person has a responsibility to contribute to the preservation and enhancement of the environment.

Sec. 102. The Congress authorizes and directs that, to the fullest extent possible: (1) the policies, regulations, and public laws of the United States shall be interpreted and administered in accordance with the policies set forth in this Act, and (2) all agencies of the Federal Government shall—

(A) Utilize a systematic, interdisciplinary approach which will insure the integrated use of the natural and social sciences and the environmental design arts in planning and in decisionmaking which may have an impact on man's environment;

(B) Identify and develop methods and procedures, in consultation with the Council on Environmental Quality established by title II of this Act, which will insure that presently unquantified environmental amenities and values may be given appropriate consideration in decisionmaking along with economic and technical considerations;

(C) Include in every recommendation or report on proposals for legislation and other major Federal actions significantly affecting the quality of the human environment, a detailed statement by the responsible official on—

(i) The environmental impact of the proposed action,

(ii) Any adverse environmental effects which cannot be avoided should the proposal be implemented,

(iii) Alternatives to the proposed action,

(iv) The relationship between local short-term uses of man's environment and the maintenance and enhancement of long-term productivity, and

(v) Any irreversible and irretrievable commitments of resources which would be involved in the proposed action should it be implemented. . . .

TITLE II: COUNCIL ON ENVIRONMENTAL QUALITY

Sec. 201. The President shall transmit to the Congress annually beginning July 1, 1970, an Environmental Quality Report. . . .

Sec. 202. There is created in the Executive Office of the President a Council on Environmental Quality (hereinafter referred to as the "Council"). The Council shall be composed of three members who shall be appointed by the President to serve at his pleasure, by and with the advice and consent of the Senate. . . .

Sec. 204. It shall be the duty and function of the Council—

(1) to assist and advise the President in the preparation of the Environmental Quality Report required by section 201 of this title;

(2) to gather timely and authoritative information concerning the conditions and trends in the quality of the environment both current and prospective, to analyze and interpret such information for the purpose of determining whether such conditions and trends are interfering, or are likely to interfere, with the achievement of the policy set forth in Title I of this Act, and to compile and submit to the President studies relating to such conditions and trends;

(3) to review and appraise the various programs and activities of the Federal Government in the light of the policy set forth in Title I of this Act for the purpose of determining the extent to which such programs and activities are contributing to the achievement of such policy, and to make recommendations to the President with respect thereto;

(4) to develop and recommend to the President national policies to foster and promote the improvement of environmental quality to meet the conservation, social, economic, health, and other requirements and goals of the Nation;

(5) to conduct investigations, studies, surveys, research, and analyses relating to ecological systems and environmental quality;

(6) to document and define changes in the natural environment, including the plant and animal systems, and to accumulate necessary data and other information for a continuing analysis of these changes or trends and an interpretation of their underlying causes;

(7) to report at least once each year to the President on the state and condition of the environment; and

(8) to make and furnish such studies, reports thereon, and recommendations with respect to matters of policy and legislation as the President may request.

Document 21
NOISE CONTROL ACT OF 1972, U.S.C. 4901–4918

The Noise Pollution and Abatement Act was enacted in 1970, but amended and replaced by the Noise Control Act of 1972. The act empowers the Environmental Protection Agency to set standards and to delegate enforcement to individual states through EPA-approved programs. As mentioned in Chapter 1, under the Noise Control Act of 1972, the primary duty of noise control rests with state and local governments, but the federal government can take action where national uniformity is required because the product will be used between states and it is a major noise source. The act forbids the state from developing standards that are more strict than the federal standard.

FINDINGS AND POLICY

Sec. 2. (a) The Congress finds—

(1) that inadequately controlled noise presents a growing danger to the health and welfare of the Nation's population, particularly in urban areas;

(2) that the major sources of noise include transportation vehicles and equipment, machinery, appliances, and other products in commerce; and

(3) that, while primary responsibility for control of noise rests with State and local governments, Federal action is essential to deal with major noise sources in commerce control of which require national uniformity of treatment.

(b) The Congress declares that it is the policy of the United States to promote an environment or all Americans free from noise that jeopardizes their health or welfare. To that end, it is the purpose of the Act to establish a means for effective coordination of Federal research and activities in noise control, to authorize the establishment of Federal noise emission standards for products distributed in commerce, and to provide information to the public respecting the noise emission and noise reduction characteristics of such products. . . .

IDENTIFICATION OF MAJOR NOISE SOURCES; NOISE CRITERIA AND CONTROL TECHNOLOGY

Sec. 5. (a)(1) The Administrator shall, after consultation with appropriate Federal agencies and within nine months of the date of the enactment of this Act, develop and publish criteria with respect to noise. Such criteria shall reflect the scientific knowledge most useful in indicating the kind and extent of all identifiable effects on the public health or welfare which may be expected from differing quantities and qualities of noise.

ENFORCEMENT

Sec. 11. (a)(1) Any person who willfully or knowingly violates . . . this Act shall be punished by a fine of not more than $25,000 per day of violation, or by imprisonment for not more than one year, or by both. . . .

CITIZEN SUITS

Sec. 12. (a) Except as provided in subsection (b), any person (other than the United States) may commence a civil action on his own behalf.

Document 22
THE SAFE DRINKING WATER ACT, 42 U.S.C. S/S 300f
ET SEQ. (1974)

The Safe Drinking Water Act was enacted by Congress in 1974, and amended in 1986 and 1996. The Act provides for the safety of drinking water supplies in the United States by establishing and enforcing national drinking water quality standards. The Act also provides for the establishment of primary regulations governing public water supplies for the protection of public health and secondary regulations regarding the taste, odor, and appearance of drinking water. The Act requires water supply system operators to use the best available technology that is economically achievable and technologically feasible. The EPA has the primary duty to set the national drinking water quality standards, to review and approve applications from the various states to assume primacy in the enforcement of those standards, and to supervise public water supply systems and other sources of drinking water. Another significant provision of the Act is that it sets up a state revolving fund system to provide money to communities to improve their drinking water facilities.

PART A—DEFINITIONS

Sec. 1401. For purposes of this title:

(1) The term "primary drinking water regulation" means a regulation which—

(A) applies to public water systems;

(B) specifies contaminants which, in the judgment of the Administrator, may have any adverse effect on the health of persons;

(C) specifies for each such contaminant either—

(i) a maximum contaminant level, if, in the judgment of the Administrator, it is economically and technologically feasible to ascertain the level of such contaminant in water in public water systems, or

(ii) if, in the judgment of the Administrator, it is not economically or technologically feasible to so ascertain the level of such contaminant, each treatment technique known to the Administrator which leads to a reduction in the level of such contaminant sufficient to satisfy the requirements of section 1412; and

(D) contains criteria and procedures to assure a supply of drinking water which dependably complies with such maximum contaminant levels; . . .

(2) The term "secondary drinking water regulation" means a regulation which applies to public water systems and which specifies the maximum

contaminant levels which, in the judgment of the Administrator, are requisite to protect the public welfare. . . .

STATE PRIMARY ENFORCEMENT RESPONSIBILITY

Sec. 1413. (a) For purposes of this title, a State has primary enforcement responsibility for public water systems during any period for which the Administrator determines (pursuant to regulations prescribed under subsection (b)) that such State—

(1) has adopted drinking water regulations which are no less stringent than the national primary drinking water regulations in effect under such sections 1412 (a) and 1412 (b);

(2) has adopted and is implementing adequate procedures for the enforcement of such State regulations, including conducting such monitoring and making such inspections as the Administrator may require by regulation;

Sec. 1417. PROHIBITION ON USE OF LEAD PIPES, SOLDER, AND FLUX.

(a) IN GENERAL.—

(1) PROHIBITION.—Any pipe, solder, or flux, which is used after the enactment of the Safe Drinking Water Act Amendments of 1986, in the installation or repair of—

(A) any public water system, or

(B) any plumbing in a residential or nonresidential facility providing water for human consumption which is connected to a public water system, shall be lead free (within the meaning of subsection (d)). This paragraph shall not apply to leaded joints necessary for the repair of cast iron pipes. . . .

CITIZEN'S CIVIL ACTION

Sec. 1449. (a) Except as provided in subsection (b) of this section, any person may commence a civil action on his behalf—

(1) against any person (including (A) the United States, and (B) any other governmental instrumentality or agency to the extent permitted by the eleventh amendment to the Constitution) who is alleged to be in violation of any requirement prescribed by or under this title, or

(2) against the Administrator where there is alleged a failure of the Administrator to perform any act or duty under this title which is not discretionary with the Administrator.

Document 23
THE TOXIC SUBSTANCE CONTROL ACT, 15 U.S.C.
2601–2671

There are over seventy thousand commercial chemicals used in the United States, with unknown toxic or dangerous characteristics. In order to prevent tragic consequences, the Toxic Substance Control Act (TSCA) requires that the EPA maintain a comprehensive inventory of all these chemicals. The Act was enacted in 1976 and amended three times since then. Under TSCA and its amendments, Congress established a number of new requirements and authorities for identifying and controlling toxic chemical hazards to human health or the environment. A chief goal of TSCA is to control chemical risks before a substance has been introduced into commerce. To accomplish this goal, mechanisms have now been developed to gather information about the toxicity of particular exposure, as well as to evaluate whether or not they cause unreasonable risk to humans and the environment. New testing standards for toxic chemicals have been developed, and appropriate control-actions have been instituted for specific chemicals. TSCA directs the EPA before undertaking regulatory actions to balance the chemical's identified risk against benefits to society and the economy. Thus, economic aspects are considered before banning toxic chemicals.

Sec. 2. FINDINGS, POLICY, AND INTENT.
(a) FINDINGS.—The Congress finds that—

(1) human beings and the environment are being exposed each year to a large number of chemical substances and mixtures.

(2) among the many chemical substances and mixtures which are constantly being developed and produced, there are some whose manufacture, processing, distribution in commerce, use, or disposal may present an unreasonable risk of injury to health or the environment; and

(3) the effective regulation of interstate commerce in such chemical substances and mixtures also necessitates the regulation of intrastate commerce in such chemical substances and mixtures.

(b) POLICY.—It is the policy of the United States that—

(1) adequate data should be developed with respect to the effect of chemical substances and mixtures on health and the environment and that the development of such data should be the responsibility of those who manufacture and those who process such chemical substances and mixtures;

(2) adequate authority should exist to regulate chemical substances and mixtures which present an unreasonable risk of injury to health or the envi-

ronment, and to take action with respect to chemical substances and mixtures which are imminent hazards; and

(3) authority over chemical substances and mixtures should be exercised in such a manner as not to impede unduly or create unnecessary economic barriers to technological innovation while fulfilling the primary purpose of this Act to assure that such innovation and commerce in such chemical substances and mixtures do not present an unreasonable risk of injury to health or the environment.

(c) INTENT OF CONGRESS.—It is the intent of Congress that the Administrator shall carry out this Act in a reasonable and prudent manner, and that the Administrator shall consider the environmental, economic, and social impact of any action the Administrator takes or proposes to take under this Act. . . .

TITLE III—INDOOR RADON ABATEMENT

Sec. 301. NATIONAL GOAL.

The national long-term goal of the United States with respect to radon levels in buildings is that the air within buildings in the United States should be as free of radon as the ambient air outside of buildings.

Document 24
THE COMPREHENSIVE ENVIRONMENTAL RESPONSE, COMPENSATION, AND LIABILITY ACT, 42 U.S.C. S/S 9601 ET SEQ. (1980)

Following well-publicized incidents, such as what occurred at Love Canal near Niagara Falls, New York, caused by the uncontrolled and dangerous disposal of hazardous wastes, it became clear that the regulations set by RCRA were primarily prospective rather than remedial. Thousands of uncontrolled or abandoned hazardous waste sites across the United States have been identified, and even where state and local governments had created their regulatory framework for the cleanup of these sites, they often lacked funds and the legal authority to deal adequately with the problem. The polluters responsible for the wastes frequently could not be found, and even when the parties could be located, legal liability was at times difficult to establish.

In 1980, responding to this situation, Congress passed CERCLA, which stands for the Comprehensive Environmental Response, Compensation, and Liability Act, also known as "Superfund." The Act was amended in 1986 by the Superfund Amendments and Reauthorization Act.

These laws authorize the EPA to respond to hazardous spills and clean up inactive dumps by either filing suit against the polluters responsible, issuing these parties an EPA order, or using a trust fund known as the Superfund. Under the Act, firms are strictly liable for all costs of removal or remedial action. Thus no negligence needs to be shown. If the EPA must conduct the cleanup because the polluters responsible for creating the hazards are not willing or able to do this, the government may seek reimbursement by the polluters under the cost recovery provision of the statute. The polluters responsible can be liable for punitive damages up to triple the cleanup costs.

These statutes established a $9 billion fund to cover the costs of the cleanup of inactive hazardous dump sites. Taxes on business finance most of the fund. However, because funding is inadequate, the Act is not intended to provide a remedy for private citizens injured by abandoned hazardous waste sites. Injured victims must file a lawsuit under the common law principles. However, if the defendant is unknown, the injured person may make limited claim against the Superfund.

REPORTABLE QUANTITIES AND ADDITIONAL DESIGNATIONS

Sec. 102. (a) The Administrator shall promulgate and revise as may be appropriate, regulations designating as hazardous substances . . . such elements, compounds, mixtures, solutions, and substances which, when released into the environment may present substantial danger to the public health or welfare or the environment, and shall promulgate regulations establishing the quantity of any hazardous substance the release of which shall be reported pursuant to section 103 of this title. . . .

CIVIL PENALTIES AND AWARDS

Sec. 109. (A) CLASS I ADMINISTRATIVE PENALTY.—
(1) VIOLATIONS.—A civil penalty of not more than $25,000 per violation may be assessed by the President. . . .

(d) AWARDS.—The President may pay an award of up to $10,000 to any individual who provides information leading to the arrest and conviction of any person for a violation subject to a criminal penalty under this Act, including any violation of section 103 and any other violation referred to in this section. . . .

USES OF FUND

Sec. 111. (A) IN GENERAL.—For purposes specified in this section there is authorized to be appropriated from the Hazardous Substance Superfund established under subchapter A of chapter 98 of the Internal Revenue Code of 1986 not more than $8,500,000,000 for the 5-year period beginning on the date of enactment of the Superfund Amendments and Reauthorization Act of 1986, and not more than $5,100,000,000 for the period commencing October 1, 1991, and ending September 30, 1994, and such sums shall remain available until expended. . . .

MAJOR INTERNATIONAL AGREEMENTS IN THE FIELD OF THE ENVIRONMENT

During the twentieth century, environmental challenges have been a priority on the agenda for international cooperation. The extent of formal international cooperation is illustrated by the growing number of UN-sponsored global conferences that explored environmental concerns (see Table 2). These conferences have expanded the involvement of international organizations in environmental issues (see Table 3). The conferences have produced over 170 international agreements in the field of the environment (see Table 4). Almost two-thirds of these agreements were signed after the 1970s. In addition, many other regional and bilateral agreements have addressed environmental issues indirectly. Taken together, all these international agreements today help manage environmental problems that cross national borders.

Under these agreements, political jurisdictions give up a degree of sovereignty to achieve mutually beneficial objectives. Though the effectiveness of particular law or agreement can be difficult to measure, the fact that nations have agreed to cooperate in so many ways suggests that efforts to protect the environment have evolved substantially during the twentieth century.

These international cooperations most often take the form of conventions, treaties, or agreements, but they sometimes are concluded memoranda of understanding arrangements, agreed measures, exchanges of letters, resolutions, or minutes. A subsequent or subsidiary agreement that originates from an existing agreement, but which itself is of substantial importance, is often called a protocol. A partial chronological list of excerpts from the major agreements follows.

Table 2
Environmental Conferences Sponsored by the United Nations, 1974–1997

Date	Conference	Location
1974	World Population Conference	Bucharest, Romania
	World Food Conference	Rome, Italy

Table 2 (continued)

Date	Conference	Location
1976	Human Settlements Conference	Vancouver, Canada
1977	UNESCO/UNEP Intergovernmental Conference on Environmental Education	Tbilisi, USSR
	U.N. Water Conference	Mar del Plata, Argentina
	U.N. Conference on Desertification	Nairobi, Kenya
	World Conference on the Ozone Layer	Washington, D.C.
1979	Conference on Science and Technology for Development	Vienna, Austria
1981	U.N. Environmental Program Senior Officials Meeting on Environmental Law	Montevideo, Uruguay
	Conference on New and Renewable Sources of Energy	Nairobi, Kenya
1982	U.N. Environment Program Session of Special Character	Nairobi, Kenya
1984	U.N. International Conference on Population	Mexico City, Mexico
1985	Vienna Convention on the Protection of the Ozone Layer Diplomatic Conference	Vienna, Austria
1987	Montreal Protocol Diplomatic Conference Protection of the Ozone Layer	Montreal, Canada
1988	Intergovernmental Panel on Climate Change	Geneva, Switzerland
1989	Basel Convention on Transboundary Movement of Hazardous Wastes Diplomatic Conference	Basel, Switzerland
1992	Conference on Environment and Development	Rio de Janeiro, Brazil
1993	Conference on Straddling Fish Stocks and Highly Migratory Fish	New York, USA
1995	Conference of the Parties to the U.N. Framework Convention on Climate Change	New York, USA
1997	Kyoto Protocol Diplomatic Conference on Controlling Gases that Contribute to Global Warming	Kyoto, Japan

Table 3
International Organizations Involved with Environmental Issues

Agency (Year Established)	Environmental Activities
UNITED NATIONS ENTITIES:	
Economic Commission for Europe (1947)	Air and water pollution control; urban development.
Intergovernmental Maritime Consultative Organization (1948)	Information on oil spill control; development of international controls over marine pollution.
U.N. Educational, Scientific, and Cultural Organization (1946)	Scientific studies on changes in oceans; development of Man and the Biosphere Program.
U.N. Development Program (1965)	Resource surveys, conservation projects, ecological studies, training programs.
U.N. Scientific Committee on the Effects of Atomic Radiation (1955)	Technical reports on levels of natural and man-made radioactivity in the environment.
International Labor Organization (1919)	Studies and standards for occupational health.
World Health Organization (1946)	Definition of environmental standards for human health; identification of environmental hazards—air, water, soil, and food pollution.
Food and Agricultural Organization (1945)	Conservation aspects of soil, forests, and terrestrial waters; studies on water quality criteria for fish; pulp and water mill effluents; sewage effluents.
World Meteorological Organization (1873)	World weather watch information related to air pollution and maritime pollution.
International Civil Aviation Organization (1947)	Aircraft noise.
International Atomic Energy Agency (1957)	Pollution from radioactive substances.
U.N. Conference of the Human Environment (1972)	Preparatory work for the 1972 Stockholm Conference.
World Bank Group (1945)	International Bank Reconstruction and Development, International Development Association, International Finance Corporation.
Regional Development Banks	African Development Bank, InterAmerican Development Bank, Asian Development Bank, Caribbean Development Bank.

Table 3 (continued)

Agency (Year Established)	Environmental Activities
U.N. Environment Programme (1972)	Of all international organizations concerned with the environment, this entity has the greatest multifaceted mission. Its mandate is to assess, monitor, and protect the human environment by seeking solutions to pollution and man-made contamination, as well as to promote environmentally sound economic and social development in both rural and urban areas.
World Commission on Environment and Development (1983)	Examines the relationship between environment and development.
OTHER INTERGOVERNMENTAL BODIES:	
NATO Committee on Challenges of Modern Society (1969)	Disaster assistance, air pollution, open-water pollution, inland pollution, environment in the strategy of regional development
Organization for Economic Cooperation and Development (1961)	Water resources management, air pollution, pesticides, organizing an environmental division
Organization for African Unity (1963)	Serves as the depository for the African Convention on the Conservation of Nature and Natural Resources
Organization of American States (1890)	Latin American Conservation on Nature Protection and wildlife preservation in the Western Hemisphere
Council of Europe (1949)	Information sharing pollution control, ecology
European Communities (1967)	European Economic and Social Cooperation
Intergovernmental Panel on Climate Change (1988)	Examines the science of climate change, its economic and social impacts, and possible response options
NONGOVERNMENTAL ORGANIZATIONS	
International Council of Scientific Unions (1919)	International Biological Program
International Union for Conservation of Nature and Natural Resources (1948)	Ecology, species survival, national parks, education, environmental policy
The World Wildlife Fund (1961)	Ecological consequences of international development

Table 4
Major International Treaties and Other Agreements in the Field of the Environment

---

Geneva (1921): Convention Concerning the Use of White Lead in Painting

To protect workers from exposure to white lead and sulfate lead from all products containing these pigments.

London (1933): Convention Relative to the Preservation of Fauna and Flora in Their Natural State

To preserve the natural fauna and flora of certain parts of the world, particularly of Africa, by means of national parks and reserves, and by regulation of hunting and collecting of species.

Washington (1946): International Convention for the Regulation of Whaling (as amended)

To protect all species of whales from overfishing and safeguard, for future generations, the great natural resource represented by whale stocks; to establish a system of international regulation for the whale fisheries in order to maintain proper conservation and development of whale stocks.

London (1954): International Convention for the Protection of Pollution of the Sea by Oil (as amended on 11 April 1962 and 21 October 1969)

To take action to prevent pollution of the sea by oil discharged from ships.

Geneva (1960): Convention Concerning the Protection of Workers Against Ionizing Radiations

To protect workers, as regards to their health and safety, against ionizing radiation.

Moscow (1963): Treaty Banning Nuclear Weapon Tests in the Atmosphere, in Outer Space and under Water

To obtain an agreement on general and complete disarmament under strict international control in accordance with the objectives of the United Nations; to put an end to the armaments race and eliminate incentives to the production and testing of all kinds of weapons, including nuclear weapons.

Washington (1973): Convention on International Trade in Endangered Species of Wild Fauna and Flora

To protect certain endangered species from overexploitation by means of a system of import/export permits.

Geneva (1979): Convention on Long-Range Transboundary Air Pollution

To protect man and his environment against air pollution and to endeavor to limit and, as far as possible, gradually reduce and prevent air pollution, including long-range transboundary air pollution.

Montego Bay (1982): United Nations Convention on the Law of the Sea

To set up a comprehensive new legal regime for the seas and oceans and, as far as environmental provisions are concerned, to establish material rules concerning environmental standards as well as enforcement provisions dealing with pollution of the marine environment.

Montreal (1987): Montreal Protocol on the Substances That Deplete the Ozone Layer

To protect the ozone layer by taking precautionary measures to control global emissions of substances that deplete it.

Rio de Janeiro (1992): Framework Convention on Climate Change

Commits states to the aim of reducing emissions of greenhouse gases.

Rio de Janeiro (1992): Convention on Biodiversity

The convention seeks to conserve biodiversity through provisions that encourage nations to take domestic actions to conserve biodiversity, promote sustainable use of biodiversity, promote benefit sharing through international cooperation, and participate in a global forum on biodiversity.

Paris (1994): Convention on Desertification

Fights against desertification by implementing strategies focused on sustainable development of land and water.

Kyoto (1997): Kyoto Protocol on Greenhouse Gases

Meeting reaches accord to reduce greenhouse gases that scientists say are warming the Earth's atmosphere.

In addition to those listed, many other regional and bilateral agreements have addressed environmental concerns indirectly.

Document 25
DECLARATION OF THE UNITED NATIONS
CONFERENCE ON THE HUMAN ENVIRONMENT

The 1972 Stockholm U.N. Conference on the Human Environment was the first high-level intergovernmental conference addressing a range of environmental concerns. Its organizers established both the processes and the focus of future conferences.

The United Nations Conference on the Human Environment,
Having met at Stockholm from 5 to 16 June 1972,
Having considered the need for a common outlook and for common principles to inspire and guide the peoples of the world in the preservation and enhancement of the human environment,

Proclaims that:

1. Man is both creature and moulder of his environment, which gives him physical sustenance and affords him the opportunity for intellectual, moral, social and spiritual growth. In the long and tortuous evolution of the human race on this planet a stage has been reached when, through the rapid acceleration of science and technology, man has acquired the power to transform his environment in countless ways and on an unprecedented scale. Both aspects of man's environment, the natural and the man-made, are essential to his well-being and to the enjoyment of basic human rights—even the right to life itself.

2. The protection and improvement of the human environment is a major issue which affects the well-being of peoples and economic development throughout the world; it is the urgent desire of the peoples of the whole world and the duty of all Governments. . . .

PRINCIPLE 4

Man has a special responsibility to safeguard and wisely manage the heritage of wildlife and its habitat, which are now gravely imperiled by a combination of adverse factors. Nature conservation including wildlife must therefore receive importance in planning for economic development. . . .

PRINCIPLE 21

States have, in accordance with the Charter of the United Nations and the principles of international law, the sovereign right to exploit their own resources pursuant to their own environmental policies, and the responsibility to ensure that activities within their jurisdiction or control do not cause damage to the environment of other States or of areas beyond the limits of national jurisdiction.

PRINCIPLE 22

States shall cooperate to develop further the international law regarding liability and compensation for the victims of pollution and other environmental damage caused by activities within the jurisdiction or control of such States to areas beyond their jurisdiction.

Source: Rowland, Wade. *The Plot to Save the World* (Toronto: Clarke, Irwin & Co., 1973), pp. 140–145.

Document 26
CONVENTION ON INTERNATIONAL TRADE IN ENDANGERED SPECIES OF WILD FAUNA AND FLORA (CITES)

Signed in 1973 and amended in 1979 and 1987, CITES is the cornerstone of international efforts to deal with threatened and endangered species. The Convention is designated to authorize the nonexploitation of wild fauna and flora. CITES divides species to be regulated into three categories (listed as Appendix I, II, III, respectively).

The Contracting States,

Recognizing that wild fauna and flora in their many beautiful and varied forms are an irreplaceable part of the natural systems of the earth which must be protected for this and the generations to come;

Conscious of the ever-growing value of wild fauna and flora from aesthetic, scientific, cultural, recreational and economic points of view;

Recognizing, in addition, that international cooperation is essential for the protection of certain species of wild fauna and flora against overexploitation through international trade;

Convinced of the urgency of taking appropriate measures to this end;

Have agreed as follows:

ARTICLE II: FUNDAMENTAL PRINCIPLES

1. . . . Appendix I shall include all species threatened with extinction which are or may be affected by trade. Trade in specimens of these species must be subject to particularly strict regulation in order not to endanger further their survival and must only be authorized in exceptional circumstances.

2. . . . Appendix II shall include:

(a) . . . all species which although not necessarily now threatened with extinction may become so unless trade in specimens of such species is subject to strict regulation in order to avoid utilization incompatible with their survival; and

(b) . . . other species which must be subject to regulation in order that trade in specimens of certain species referred to in subparagraph (a) of this paragraph may be brought under effective control.

3. . . . Appendix III shall include all species which any Party identifies as being subject to regulation within its jurisdiction for the purpose of preventing or restricting exploitation, and as needing the cooperation of other Parties in the control of trade.

4.... The Parties shall not allow trade in specimens of species included in Appendices I, II and III except in accordance with the provisions of the present Convention.

ARTICLE III: REGULATION OF TRADE IN SPECIMENS
OF SPECIES

2.... The export of any specimen of a species [included in Appendices I, II, and III] shall require the prior grant and presentation of an export permit. An export permit shall only be granted when the following conditions have been met:

(a) ... a Scientific Authority of the State of export has advised that such export will not be detrimental to the survival of that species;

(b) ... a Management Authority of the State of export is satisfied that the specimen was not obtained in contravention of the laws of that State for the protection of fauna and flora.

Source: UNEP Register of International Treaties and other Agreements in the Field of the Environment.

Document 27
MONTREAL PROTOCOL ON SUBSTANCES THAT
DEPLETE THE OZONE LAYER

Signed in 1987, and amended in 1990, 1992, and 1995, the principal goal of the Montreal Protocol is to limit and reduce the use of chemicals that destroy the stratospheric ozone layer. Each chemical with a scientifically determined ozone depletion level is listed in Table 3.1. The depletion level is used in determining a control level, from which future reductions in use of the chemical are calculated.

The Parties to this Protocol,

Being Parties to the Vienna Convention for the Protection of the Ozone Layer,

Mindful of their obligation under that Convention to take appropriate measures to protect human health and the environment against adverse effects resulting or likely to result from human activities which modify or are likely to modify the ozone layer,

Recognizing that world-wide emissions of certain substances can significantly deplete and otherwise modify the ozone layer in a manner that is likely to result in adverse effects on human health and the environment,

Conscious of the potential climatic effect of emissions of these substances,

Aware that measures taken to protect the ozone layer from depletion should be based on relevant scientific knowledge, taking into account technical and economic considerations,

Determined to protect the ozone layer by taking precautionary measures to control equitably total global emissions of substances that deplete it, with the ultimate objective of their elimination on the basis of developments in scientific knowledge, taking into account technical and economic considerations and bearing in mind the developmental needs of developing countries,

Acknowledging that special provision is required to meet the needs of developing countries, including the provision of additional financial resources and access to relevant technologies, bearing in mind that the magnitude of funds necessary is predictable, and the funds can be expected to make a substantial difference in the world's ability to address the scientifically established problem of ozone depletion and its harmful effects,

Noting the precautionary measures for controlling emissions of certain chlorofluorocarbons that have already been taken at national and regional levels,

Considering the importance of promoting international cooperation in the research, development and transfer of alternative technologies relating to the control and reduction of emissions of substances that deplete the ozone layer, bearing in mind in particular the needs of developing countries,

Have agreed as follows:

ARTICLE 2: CONTROL MEASURES

1. Each Party shall ensure that for the twelve-month period commencing on the first day of the seventh month following the date of the entry into force of this Protocol, and in each twelve-month period thereafter, its calculated level of consumption of the controlled substances [listed in Table 3.1 . . . does not exceed its calculated level of consumption in 1986.

ARTICLE 11: MEETING OF THE PARTIES

1. The Parties shall hold meetings at regular intervals. The secretariat shall convene the first meeting of the Parties not later than one year after the date of the entry into force of this Protocol and in conjunction with a meeting of the Conference of the Parties to the Convention, if a meeting of the latter is scheduled within that period.

ARTICLE 16: ENTRY INTO FORCE

1. This Protocol shall enter into force on 1 January 1989.

Source: UNEP Register of International Treaties and other Agreements in the Field of the Environment.

Document 28
ENVIRONMENTAL DECLARATIONS FROM THE ECONOMIC SUMMIT, PARIS 1989

The yearly summit of seven industrialized nations, called the "G-7," includes the political leaders of Canada, Germany, France, Italy, Japan, the United Kingdom, and the United States. Summits in the 1980s began to give attention to environmental concerns. In its 1989 annual meeting, environmental issues constituted one-third of the Paris Summit Declaration. The following excerpts from the G-7 meeting declarations show the environmental concerns of the industrialized nations represented at the Paris meeting.

33) There is a growing awareness throughout the world of the necessity to preserve better the global ecological balance. This includes serious threats to the atmosphere, which could lead to future climate changes. We note with great concern the growing pollution of air, lakes, rivers, oceans, and seas;

acid rain, dangerous substances; and the rapid desertification and deforestation. Such environmental degradation endangers species and undermines the well-being of individuals and societies.

Decisive action is urgently needed to understand and protect the earth's ecological balance. We will work together to achieve the common goals of preserving a healthy and balanced global environment in order to meet shared economic and social objectives and to carry out obligations to future generations.

34) We urge all countries to give further impetus to scientific research on environmental issues, to develop necessary technologies and to make clear evaluations of the economic costs and benefits of environmental policies.

The persisting uncertainty on some of these issues should not unduly delay our action. In this connection, we ask all countries to combine their efforts in order to improve observation and monitoring on a global scale.

35) We believe that international cooperation also needs to be enhanced in the field of technology and technology transfer in order to reduce pollution or provide alternative solutions.

36) We believe that industry has a crucial role in preventing pollution at source, in waste minimization, in energy conservation, and in the design and marketing of cost-effective clean technologies. The agriculture sector must also contribute to tackling problems such as water pollution, soil erosion, and desertification.

37) Environmental protection is integral to issues such as trade, development, energy, transport, agriculture, and economic planning. Therefore, environmental considerations must be taken into account in economic decision-making. In fact good economic policies and good environmental policies are mutually reinforcing.

In order to achieve sustainable development, we shall ensure the compatibility of economic growth and development with the protection of the environment. Environmental protection and related investment should contribute to economic growth. In this respect, intensified efforts for technological breakthrough are important to reconcile economic growth and environmental policies.

Clear assessments of the costs, benefits, and resource implications of environmental protection should help governments to make the necessary decisions on the mix of price signals (e.g., taxes or expenditures) and regulatory actions, reflecting where possible the full value of natural resources.

We encourage the World Bank and the regional development banks to integrate environmental considerations into their activities. International organizations such as the OECD and the United Nations and its affiliated organizations, will be asked to develop further techniques of analysis which would help governments assess appropriate economic measures to promote the quality of the environment. We ask the OECD, within the context of its work on integrating environment and economic decisionmaking, to examine how selected environmental indicators could be developed. We expect the 1992 U.N. Conference on Environment and Development to give additional momentum to the protection of the global environment.

Source: Weekly Compilation of Presidential Documents, July 6, 1989.

Document 29
THE RIO DECLARATION ON ENVIRONMENT AND DEVELOPMENT

To celebrate the twentieth anniversary of the historic Stockholm Conference, the United Nations sponsored the U.N. Conference on Environment and Development. Held June 3–14, 1992, in Rio de Janeiro, Brazil, this international conference was the largest event of its kind in history. It promised to shape international environment and economic development agendas into the twenty-first century. The Rio Declaration, a nonbinding statement, sets forth 27 basic principles that will guide environmental protection and economic development worldwide.

The United Nations Conference on Environment and Development,

Having met at Rio de Janeiro from 3 to 14 June 1992,

Reaffirming the Declaration of the United Nations Conference on the Human Environment, adopted at Stockholm on 16 June 1972, and seeking to build upon it,

With the goal of establishing a new and equitable global partnership through the creation of new levels of cooperation among States, key sectors of societies and people,

Working towards international agreements which respect the interests of all and protect the integrity of the global environmental and developmental system,

Recognizing the integral and interdependent nature of the Earth, our home,

Proclaims that:

1. Human beings are the center of concerns for sustainable development. They are entitled to a healthy and productive life in harmony with nature.

2. States have, in accordance with the Charter of the United Nations and the principles of international law, the sovereign right to exploit their own resources pursuant to their own environmental and developmental policies, and the responsibility to ensure that activities within their jurisdiction or control do not cause damage to the environment of other States or of areas beyond the limits of national jurisdiction.

3. The right of development must be fulfilled so as to equitably meet developmental and environmental needs of present and future generations.

4. In order to achieve sustainable development, environmental protection shall constitute an integral part of the development process and cannot be considered in isolation from it.

5. All States and all people shall cooperate in the essential task of eradicating poverty as an indispensable requirement for sustainable development, in order to decrease the disparities in standards of living and better meet the needs of the majority of the people of the world.

6. The special situation and needs of developing countries, particularly the least developed and those most environmentally vulnerable, shall be given special priority. International actions in the field of the environment and development should also address the interests and needs of all countries.

7. States shall cooperate in a spirit of global partnership to conserve, protect and restore the health and integrity of the Earth's ecosystem. In view of the different contributions to global environmental degradation, States have common but differentiated responsibilities. The developed countries acknowledge the responsibility that they bear in the international pursuit of sustainable development in view of the pressures their societies place on the global environment and of the technologies and financial resources they command.

8. To achieve sustainable development and a higher quality of life for all people, States should reduce and eliminate unsustainable patterns of production and consumption and promote appropriate demographic policies.

9. States should cooperate to strengthen endogenous capacity-building for sustainable development by improving scientific understanding through exchanges of scientific and technological knowledge, and by enhancing the development, adaptation, diffusion and transfer of technologies, including new and innovative technologies.

10. Environmental issues are best handled with the participation of all concerned citizens, at the relevant level. At the national level, each individual shall have appropriate access to information concerning the environment that is held by public authorities, including information on hazardous materials and activities in their communities, and the opportunity to participate in decision-making processes. States shall facilitate and encourage public awareness and participation by making information widely available. Effective access to judicial and administrative proceedings, including redress and remedy, shall be provided.

11. States shall enact effective environmental legislation. Environmental standards, management objectives and priorities should reflect the environmental and developmental context to which they apply. Standards applied by some countries may be inappropriate and of unwarranted economic and social cost to other countries, in particular developing countries.

12. States should cooperate to promote a supportive and open international economic system that would lead to economic growth and sustainable development in all countries, to better address the problems of environmental degradation. Trade policy measures for environmental purposes should not constitute a means of arbitrary or unjustifiable discrimina-

tion or a disguised restriction on international trade. Unilateral actions to deal with environmental challenges outside the jurisdiction of the importing country should be avoided. Environmental measures addressing transboundary pollution or global environmental problems should, as far as possible, be based on an international consensus.

13. States shall develop national law regarding liability and compensation for the victims of pollution and other environmental damage. States shall also cooperate in an expeditious and more determined manner to develop further international law regarding liability and compensation for adverse effects of environmental damage caused by activities within their jurisdiction or control to areas beyond their jurisdiction.

14. States should effectively cooperate to discourage or prevent the relocation and transfer to other States or any activities and substances that cause severe environmental degradation or are found to be harmful to human health.

15. In order to protect the environment, the precautionary approach shall be widely applied by States according to their capabilities. Where there are threats of serious or irreversible damage, lack of full scientific certainty shall not be used as a reason for postponing cost-effective measures to prevent environmental degradation.

16. National authorities should endeavor to promote the internalization of environmental costs and the use of economic instruments, taking into account the approach that the polluter should, in principle, bear the cost of pollution, with due regard to the public interest and without distorting international trade and investment.

17. Environmental impact assessment, as a national instrument, shall be undertaken for proposed activities that are likely to have a significant adverse impact on the environment and are subject to a decision of a competent national authority.

18. States shall immediately notify other States of any natural disasters or other emergencies that are likely to produce sudden harmful effects on the environment of those States. Every effort shall be made by the international community to help States so afflicted.

19. States shall provide prior and timely notification and relevant information to potentially affected States on activities that may have a significant adverse transboundary environmental effect and shall consult with those States at an early stage and in good faith.

20. Women have a vital role in environmental management and development. Their full participation is therefore essential to achieve sustainable development.

21. The creativity, ideals and courage of the youth of the world should be mobilized to forge a global partnership in order to achieve sustainable development and ensure a better future for all.

22. Indigenous people and their communities, and other local communities, have a vital role in environmental management and development because of their knowledge and traditional practices. States should recognize and duly support their identity, culture and interests and enable their effective participation in the achievement of sustainable development.

23. The environment and natural resources of people under oppression, domination and occupation shall be protected.

24. Warfare is inherently destructive of sustainable development. States shall therefore respect international law providing protection for the environment in times of armed conflict and cooperate in its further development, as necessary.

25. Peace, development and environmental protection are interdependent and indivisible.

26. States shall resolve all their environmental disputes peacefully and by appropriate means in accordance with the Charter of the United Nations.

27. States and people shall cooperate in good faith and in a spirit of partnership in the fulfillment of the principles embodied in this Declaration and in the further development of international law in the field of sustainable development.

Source: UNEP Register of International Treaties and other Agreements in the Field of the Environment.

Document 30
UNITED NATIONS FRAMEWORK CONVENTION ON CLIMATE CHANGE

The 1992 Framework Convention on Climate Change commits nations to the aim of reducing emissions of greenhouse gases and to the Montreal Protocol that seeks to phase out production of ozone-depleting chemicals. The objectives of the convention are to stabilize greenhouse gas concentrations in the atmosphere at a level that would prevent dangerous anthropogenic interference with the climate system "within a time frame sufficient to allow ecosystems to adapt naturally to climate change, to ensure that food production is not threatened, and to enable economic development to proceed in a sustainable manner." Originally signed at the Rio Summit in 1992, the negotiations were strengthened by conferences held at Berlin (1995), Geneva (1996), and Kyoto (1997).

The Parties to this Convention,

Acknowledging that change in the Earth's climate and its adverse effects are a common concern of humankind,

Concerned that human activities have been substantially increasing the atmospheric concentrations of greenhouse gases, that these increases enhance the natural greenhouse effect, and that this will result on average in an additional warming of the Earth's surface and atmosphere and may adversely affect natural ecosystems and humankind,

Noting that the largest share of historical and current global emissions of greenhouse gases has originated in developed countries, that per capita emissions in developing countries are still relatively low and that the share of global emissions originating in developing countries will grow to meet their social and development needs,

Aware of the role and importance in terrestrial and marine ecosystems of sinks and reservoirs of greenhouse gases,

Noting that there are many uncertainties in predictions of climate change, particularly with regard to the timing, magnitude and regional patterns thereof,

Acknowledging that the global nature of climate change calls for the widest possible cooperation by all countries and their participation in an effective and appropriate international response, in accordance with their common but differentiated responsibilities and respective capabilities and their social and economic conditions,

Recalling the pertinent provisions of the Declaration of the United Nations Conference on the Human Environment, adopted at Stockholm on 16 June 1972, . . .

Recognizing that steps required to understand and address climate change will be environmentally, socially and economically most effective if they are based on relevant scientific, technical and economic considerations and continually re-evaluated in the light of new findings in these areas,

Recognizing that various actions to address climate change can be justified economically in their own right and can also help in solving other environmental problems,

Recognizing also the need for developed countries to take immediate action in a flexible manner on the basis of clear priorities, as a first step towards comprehensive response strategies at the global, national and, where agreed, regional levels that take into account all greenhouse gases, with due consideration of their relative contributions to the enhancement of the greenhouse effect,

Recognizing further that low-lying and other small island countries, countries with low-lying coastal, arid and semi-arid areas or areas liable to floods, drought and desertification, and developing countries with fragile mountainous ecosystems are particularly vulnerable to the adverse effects of climate change,

Recognizing the special difficulties of those countries, especially developing countries, whose economies are particularly dependent on fossil fuel production, use and exportation, as a consequence of action taken on limiting greenhouse gas emissions,

Affirming that responses to climate change should be coordinated with social and economic development in an integrated manner with a view to avoiding adverse impacts on the latter, taking into full account the legitimate priority needs of developing countries for the achievement of sustained economic growth and the eradication of poverty,

Recognizing that all countries, especially developing countries, need access to resources required to achieve sustainable social and economic development and that, in order for developing countries to progress towards that goal, their energy consumption will need to grow taking into account the possibilities for achieving greater energy efficiency and for controlling greenhouse gas emissions in general, including through the application of new technologies on terms which make such an application economically and socially beneficial,

Determined to protect the climate system for present and future generations,

Have agreed as follows:

ARTICLE 2 OBJECTIVE

The ultimate objective of this Convention and any related legal instruments that the Conference of the parties may adopt is to achieve, in accordance with the relevant provisions of the Convention, stabilization of greenhouse gas concentrations in the atmosphere at a level that would prevent dangerous anthropogenic interference with the climate system. Such a level should be achieved within a time frame sufficient to allow ecosystems to adapt naturally to climate change, to ensure that food production is not threatened and to enable economic development to proceed in a sustainable manner.

ARTICLE 3 PRINCIPLES

In their actions to achieve the objective of the Convention and to implement its provisions, the Parties shall be guided, inter alia, by the following:

1. The Parties should protect the climate system for the benefit of present and future generations of humankind, on the basis of equity and in accordance with their common but differentiated responsibilities and respective capabilities. Accordingly, the developed country Parties should take the lead in combating climate change and the adverse effects thereof.

2. The specific needs and special circumstances of developing country Parties, especially those that are particularly vulnerable to the adverse effects of climate change, and of those Parties, especially developing country Parties, that would have to bear a disproportionate or abnormal burden under the Convention, should be given full consideration.

3. The Parties should take precautionary measures to anticipate, prevent or minimize the causes of climate change and mitigate its adverse effects. Where there are threats of serious or irreversible damage, lack of full scientific certainty should not be used as a reason for postponing such measures, taking into account that policies and measures to deal with climate change should be cost-effective so as to ensure global benefits at the lowest possible cost. To achieve this, such policies and measures should take into account different socio-economic contexts, be comprehensive, cover all relevant sources, sinks and reservoirs of greenhouse gases and adaptation, and comprise all economic sectors. Efforts to address climate change may be carried out cooperatively by interested Parties.

4. The Parties have a right to, and should, promote sustainable development. Policies and measures to protect the climate system against human-induced change should be appropriate for the specific conditions of each Party and should be integrated with national development programmes, taking into account that economic development is essential for adopting measures to address climate change.

5. The Parties should cooperate to promote a supportive and open international economic system that would lead to sustainable economic growth and development in all Parties, particularly developing country Parties, thus enabling them better to address the problems of climate change. Measures taken to combat climate change, including unilateral ones, should not constitute a means of arbitrary or unjustifiable discrimination or a disguised restriction on international trade.

ARTICLE 7 CONFERENCE OF THE PARTIES

1. A Conference of the Parties is hereby established.

2. The Conference of the Parties, as the supreme body of this Convention, shall keep under regular review the implementation of the Convention and any related legal instruments that the Conference of the Parties may

adopt, and shall make, within its mandate, the decisions necessary to promote the effective implementation of the Convention.

ARTICLE 8 SECRETARIAT

3. The Conference of the Parties, at its first session, shall designate a permanent secretariat and make arrangements for its functioning.

ARTICLE 9 SUBSIDIARY BODY FOR SCIENTIFIC AND TECHNOLOGICAL ADVICE

1. A subsidiary body for scientific and technological advice is hereby established to provide the Conference of the Parties and, as appropriate, its other subsidiary bodies with timely information and advice on scientific and technological matters relating to the Convention. This body shall be open to participation by all Parties and shall be multidisciplinary. It shall comprise government representatives competent in the relevant field of expertise. It shall report regularly to the Conference of the Parties on all aspects of its work.

ARTICLE 17 PROTOCOLS

1. The Conference of the Parties may, at any ordinary session, adopt protocols to the Convention.

Source: U.N. Framework Convention on Climate Change (United Nations, N.Y., 1992).

Glossary of Selected Terms

Abiotic: Without life; pertains to nonliving factors of the environment.

Acid rain: Rain or other precipitation containing nitric and sulfuric acids derived from air pollutants, such as industrial or automobile emissions.

Acute effect: An effect that appears shortly after exposure to a pollutant.

Adaptation: Structural, physiological, or behavioral characteristics that enhance the ability of the organism to survive and reproduce in a particular environment.

Administrative agency: A governmental unit, like the Environmental Protection Agency or Nuclear Regulatory Commission, established by the legislative or the executive branch to execute certain functions of the government; it allows governmental business to be transacted under the authority of the legislative or executive branch by units that specialize in certain tasks.

Age structure: The division of a population according to age groupings. The different age levels of a population are expressed in percentages; it is often depicted graphically as an age pyramid.

Albedo: The fraction or incident of light or electromagnetic radiation that is reflected by a body as compared to the reflectivity of a perfect reflector of the same size, shape, orientation, and distance; it is usually given as a percentage.

Algal bloom: A population explosion of photosynthetic organisms (particularly algae and blue-green algae) typically caused by nutrient enrichment in aquatic ecosystems.

Ambient air: Under the Clean Air Act, ambient air is the air outside of enclosures, such as buildings and houses.

Anthropocentric ethics: Values predicated by their importance to humans.

Anthropogenic: Generally used in the context of emissions that are produced as the result of human activities.

Anthroposystem: An artificial system consisting of humankind, domesticated organisms, and their nonliving environment, which form a self-sustainable unit.

Assimilative capacity: The ability of a natural ecosystem to receive waste or toxic materials without harmful effects and/or damage to the biota.

Benefit/Cost analysis: The technique of weighing the advantages and disadvantages of undertaking a project in order to determine its value.

Benign tumor: A noninvasive cell growth that does not damage normal tissue.

Biocide: A factor that kills organisms directly by damaging their bodies.

Biodegradable: Capable of being chemically broken down by the action of organisms. For example, many chemicals, food scraps, cotton, wool, and paper products are biodegradable; plastics and polyesters usually are not.

Biodiversity: The number and variety of different organisms in an ecosystem.

Biogeochemical cycle: The circular passage and transformation of matter, like carbon, from a living organism to the abiotic environment and back again.

Biomagnification: The concentration of a nonmetabolizable and nonexcretable substance, like DDT, in an organism's body beyond that fund in the environment.

Biomass: Total dry weight of all living matter of a particular species in a designated habitat or region.

Biosphere: The portion of the Earth and its atmosphere that contains living organisms.

Biota: All living species in a given region.

Biotic: The living factors of the environment.

Biotic community: Association of various interacting species in a common environment.

Biotic potential: The maximum potential growth rate of a population under ideal conditions.

Birth rate: The number of organisms born during a particular time period.

Bubble concept: Regulatory policy that allows all sources within a manufacturing firm to be treated as one source, as if they were encased in a bubble with only one point of emission.

Cancer: A malignant growth of cells that invades surrounding tissues, causing death or abnormal function of these tissues.

Carcinogen: Any substance or agent capable of inducing cancerous growth.

Carrying capacity: The average population size of a given species that an environment can support on a sustained basis by utilizing the available resources of that particular environment.

Catalytic converter: The device used on motor vehicle exhaust systems to convert noxious air pollutants into relatively benign substances.

Chlorinated hydrocarbons: A family of chemicals in which hydrogen atoms have been replaced by chlorine atoms, also called organochlorides. These include pesticides (DDT, aldrin, chlordane, and heptachlor) and numerous organic compounds (polychlorinated biphenyls).

Chlorofluorocarbons (CFCs): A group of inert, nontoxic, easily liquified chemicals used in refrigeration, air-conditioning, foam cushioning, and insulation. They are also used as solvents or aerosol propellants, which have been shown to deplete the ozone layer.

Clear cut: Harvesting all the trees in one region at one time. It is an activity that destroys key habitats and biodiversity. It also encourages rainfall or snowmelt runoff, erosion, sedimentation of streams and lakes, and flooding.

Climax community: A relatively stable community that maintains itself in a given area as long as there are no significant changes in the conditions of the environment.

Closed system: A system in which there is no exchange of matter or energy with its surroundings.

Common law: Law created by judicial decisions, as distinguished from law created by the enactments of legislatures.

Competition: Rivalry between organisms of the same or different species for required resources, such as food, water, or space.

Conservation: A form of anthropocentric ethics that believes natural resources must be held in trust for future generations through careful use and management.

Contaminated water: Water that contains toxic chemicals or pathogenic organisms.

Cost-benefit analysis: A method by which one calculates the monetary costs to both private parties and the general public of implementing a certain regulation and weighs it against the total costs of the estimated monetary value of the benefits received from the same regulation.

Death rate: The number of organisms that died during a particular time period.

Deep ecology: An extreme form of ecocentric ethics that sees humankind as no more important than other species.

Deforestation: The removal of forests by cutting and burning to provide land for agriculture, housing, building, or some similar venture, or by harvesting the trees for building materials or fuel.

Developed nations: Industrialized nations that have a strong economy and sophisticated technological development such as England, France, the United States, and Japan.

Developing nations: Nations that are not yet industrialized and have little or no technological development such as Ethiopia, Bangladesh, and Haiti.

Disinfectant: A substance that kills pathogenic organisms.

Dissolved oxygen: Free, uncombined oxygen (O_2) molecules dissolved in water.

Doubling time: The time it takes for a population to double in size.

Due process: A process that generally requires the judicial system to give notice of an accusation against oneself, as well as the opportunity to be heard and to defend oneself before an impartial tribunal. Moreover, the government cannot exercise arbitrary and unreasonable powers.

Dynamic equilibrium: A condition that exists in a system when matter and energy input equals the outflow of these resources over a period of time.

Ecocentrics: The belief that wildlife has natural rights that are independent of human-center values (e.g., anthropocentric values).

Ecological niche: The structural and functional role that a particular species plays in its environment, which is determined by the physiological or behavioral actions of a given species.

Ecological succession: The orderly and sequential transition from a pioneer to a climax community.

Ecologist: A scientist who studies the interaction of organisms between themselves and their abiotic environment.

Economic externality: The portion of cost not taken into account when evaluating the cost of doing business or carrying out an activity—for example the cost of environmental damage such as polluted air that results from manufacturing a product or driving a car.

Ecosphere: The sum total of life on Earth, together with the global environment and the Earth's total resources.

Ecosystem: The treatment of a community and its abiotic environment as an interacting functional unit.

Effluent charge: A tax, fee, or fine on a business enterprise for its polluting activity, usually on a per unit basis.

Electrostatic precipitator: A device installed in smokestacks to remove soot and other particulates.

Eminent domain: The power of the government to take private property for public use on the grounds of public exigency and for the public good.

Endangered species: Species in danger of extinction and whose survival is unlikely if the causal factors continue to operate.

Energy: The capacity to do work or to cause specific change. Energy holds matter together and can become mass or can be derived from mass.

Energy recovery: The obtaining of heat from organic wastes, as in refuse-derived fuel.

Energy source: That from which energy can be derived.

Environmental justice: The fair treatment of individuals, groups, or communities regardless of race, ethnicity, or economic status with respect to the development and enforcement of environmental laws.

Environmental resistance: Sum total of factors, such as predation and competition, in the external environment that limit the numerical increase of a population in a particular area.

Environmental scientist: A scientist that applies scientific facts and theories to solve definite environmental problems.

Environmentalist: A person who supports environmental issues above and beyond personal gains.

Equilibrium: A stable condition in which net change does not occur, because opposing changes occur at exactly balancing rates. *See* **dynamic equilibrium**.

Estuary: A coastal body of water (such as a river mouth or coastal bay), partly enclosed by land, but having free access to the open ocean. These brackish regions include bays, mouths of rivers, salt marshes, wetlands, and lagoons.

Evolution: The change over time of the proportions of individuals differing genetically in one or more traits. *See* **natural selection**.

Exponential growth: Population growth in which the total number increases in the same manner as compounded interest, which is 1, 2, 4, 8, 16, and so on.

Externality: The effect of a resource allocation decision on an individual or firm not directly participating in the decision.

Extinction: The progressive deaths of all members of a particular species.

Fauna: The animals inhabiting a given area.

Feedback: *See* **negative feedback** and **positive feedback**.

Flora: The plants inhabiting a given area.

Food chain: A linear sequence of food levels, each of which feeds on the previous one in order to acquire matter and energy.

Fossil fuels: The remains of once-living species, which are burned to release energy, such as coal, petroleum, or natural gas.

Gaia hypothesis: The hypothesis that the biosphere is a single complex superorganism that is alive, in the sense that it has a homeostatic quality and is continually able to reconstitute and repair itself. The term stems from the name of the ancient Greek goddess Gaia, the goddess of Earth.

Greenhouse effect: The heating of the atmosphere by virtue of the fact that short-wavelength solar radiation is transmitted rather freely through the atmosphere and infrared radiation from the Earth is more readily absorbed.

Greenhouse gases: Those gases, such as water vapor, carbon dioxide, tropospheric ozone, nitrous oxide, and methane, that are relatively transparent to shortwave solar radiation but opaque to longwave (infrared) radiation.

Groundwater: Water found under the ground, usually in porous rock formations.

Growth rate: The change in population size over time.

Halons (bromofluorocarbons): Nontoxic, inert chemicals that have at least one bromine atom in their chemical composition. They evaporate without leaving a chemical residue and are used in fire extinguishers and fire suppressant systems.

Homeostatis: The capacity of a system to maintain a constant internal environment while external conditions vary.

Hydrochlorofluorocarbons (HCFCs): Chemicals composed of one or more carbon atoms and varying numbers of hydrogen, chlorine, and fluorine atoms.

Instability: A property of any system that with certain disturbances or perturbations will increase in magnitude.

Intrinsic rate of increase: The rate at which a population of a particular species is capable of increasing under optimum conditions.

Landfill: A land disposal site often located without regard to possible effects on water resources, but which employs intermittent or daily cover to minimize scavenger, aesthetic, vector, and air pollution problems.

Methane: A hydrocarbon gas (CH_4) that is the principal component of natural gas.

Montreal Protocol: A treaty, signed in 1987 by most of the developed nations, to substantially reduce the use of chlorofluorocarbons (CFCs).

Mortality rate: The rate at which members of a population die. In the case of the human population, the rate is normally given in deaths per thousand of persons per annum.

Mutagen: An agent that is capable of increasing the rate of mutation in organisms.

Mutation: Change in genetic constitution of a species due to errors in the duplication of genes during meiosis. Mutations can be induced by chemicals, radiation, or a virus.

Mutualism: A form of interaction in which both organisms benefit.

Natality: The rate at which new individuals are produced in a population.

National Environmental Policy Act (NEPA): A Magna Carta-like federal program (1969) on environmental protection and management. This statute set forth the first comprehensive national environmental policy ever articulated by Congress. NEPA is divided into two titles. Title I established general policy objectives and delegated specific duties on federal agencies. This title addresses the interrelationship between all components of the natural environment including population, urbanization, industrial and technological advances, and resource exploitation. To aid in the implementation of NEPA's policy, Title II established the Council on Environmental Quality (CEQ) in the Executive Office of the U.S. President. The statute directs the CEQ to assume responsibility for advising, coordinating, and monitoring agency implementation of this new law.

Natural resources: Energy, food, and materials made available by geological processes.

Natural selection: Biological evolution by differential reproduction. Organisms best adapted to their environment will survive and will reproduce in the greatest numbers. *See* **evolution.**

Negative feedback: Control mechanism whereby an increase in concentration of some output leads to a decrease in its production and vice versa.

Negligence: In law, the omission to do something that a reasonable person, guided by those considerations that ordinarily regulate human affairs, would do, or the act of doing something that a reasonable and prudent person would not do.

Nonpoint sources: Under the Clean Air Act, these are sources of pollution that enter surface water and ground water in a diffused manner over a large area. Examples are agricultural and urban runoff from rainstorms.

Open system: A system in which constituents can move in and out of the system.

Organism: Any living being, whether bacterium, protozoan, fungus, plant, or animal.

Ozone: A molecule made up of three atoms of oxygen (O_3). Ozone is gaseous, is almost colorless (but faintly blue), and has an odor similar to weak chlorine. Ozone shields the Earth's surface from ultraviolet radiation.

Pesticides: Chemical compounds used to control organisms, such as roaches and mosquitoes, that are deemed to be pests.

Photochemical: A chemical reaction in which light is an agent or factor.

Photosynthesis: The process by which plants and most autotrophs convert carbon dioxide and water into carbohydrates through the use of solar energy. Oxygen and water are released in the process.

Point sources: Under the Clean Water Act, these are discrete sources of pollution that are easy to identify, such as wells, ditches, or pipes.

Pollution: The condition caused by the addition of sufficient materials or waste to the water, air, or land so as to reduce or destroy their normal integrity. Pollution can also be caused by the overexploitation of natural resources.

Population: A group of organisms of the same species inhabiting a specified locality.

Positive feedback: A process whereby an increase in concentration of some output leads to an increase in its production. Unless dampened by some other process, positive feedback tends to be unstable.

Potable water: Water free from harmful and unpleasant substances; water that is not dangerous to human health.

Preservationist: A form of ecocentric ethics in which use of a wilderness area is limited to nonconsumptive activities.

Probability: A mathematical expression of the degree of confidence that certain events will or will not occur.

Procedural law: The law that governs the method of enforcing rights or obtaining redress for invasion; it is a type of law that prescribes the procedure to be followed in a case.

Producers: In natural ecosystems, organisms, like plants, that are able to synthesize their own food from inorganic substances. In anthroposystems, farms and industries would be characterized as producers.

Public goods: Goods for which two or more individuals can consume the same unit at the same time and not diminish the amount available to others. For example, one person viewing a charming snow-covered mountain peak does not affect any other person's view of this charming landscape.

Recycling: The reprocessing of wastes to recover the original raw material. For example, old aluminum cans are usually melted and recast into new cans and fiber is recovered from wastepaper.

Renewable energy: Energy obtained from sources that are fundamentally inexhaustible. Renewable sources of energy encompass geothermal, hydroelectric, tidal power, wind, biofuels, solar collectors, and solar cells.

Resource recovery: Productive use of material that would otherwise be disposed of as waste. It includes recycling, material conversion, and energy recovery.

Respiration: Chemical oxidative process whereby a living organism breaks down certain organic matter with the release of energy used in metabolism.

Reuse: Object is used again in the same form. For example, cleaning glass bottles and refilling them with soda.

Second Law of Thermodynamics: Transformation of energy is always less than 100 percent efficient; when energy is changed from one form to another, some of it is lost as waste heat.

Species: Genetically, a group of actually or potentially inbreeding organisms that are reproductively isolated from other such groups.

Steady-State: An adjective describing a system that is in a stable dynamic state in which inputs balance outputs.

Stratosphere: The layer of the upper atmosphere extending from the tropopause (eight to fifteen kilometers altitude) to approximately fifty kilometers. Its thermal structure, which is determined by its radiation balance, is normally very stable with a low humidity.

Substantive law: The branch of law that prescribes legal rights as opposed to the part that prescribes methods of enforcing rights or obtaining redress for their decisions.

Succession: *See* **ecological succession.**

Sustainable development: Environmentally benign methods of development that allow the production of goods and services without damage to the ecosystem. No significant effects on soil, water supplies, biodiversity, or other surrounding natural resources. The notion of sustainability is an "intergenerational" concern in which the present generation passes on a conserved or improved natural resource base rather than one that has been depleted or polluted. Sustainable development is normally associated with farms, ranches, forestry, fisheries, and the like, that are self-sustaining.

Synergistic: The effect produced by two or more agents working simultaneously, greater than the result produced from the agents working separately.

Synstem: Any interacting, interdependent, or associated group of objects in space and/or time.

Threatened species: A species likely to become endangered in the foreseeable future.

Toxic substances: A chemical or mixture of chemicals that is poisonous to organisms.

Troposphere: The region of the lower atmosphere extending from the surface to about fifteen kilometers, within which there is generally a steady decrease of temperature with increasing altitude. Almost all clouds develop in this region.

Ultraviolet: Portion of the electromagnetic spectrum that has frequencies somewhat higher than the violet end of the visible spectrum.

Variable: Any of the characteristics that vary or change with time, such as species diversity, temperature, relative humidity, and light.

Waste-water treatment plant: A series of tanks, screens, filters, and other processes by which pollutants are removed from water.

Wetlands: Regions normally soaked or flooded by water; the adapted ecosystem subsequently characterized by the prevalence of vegetation adapted for life in saturated-soil conditions.

Annotated Bibliography

BOOKS

Adam, William M. *Green Development: Environment and Sustainability in the Third World*. London: Routledge, 1990. Provides an overview of sustainable development, poverty, and environmental activism in the developing countries.

Anderson, Terry L., and Donald R. Leal. *Free Market Environmentalism*. Boulder, CO: Westview Press, 1991. Starts with a treatment of free-market environmentalism and challenges its limits. It emphasizes the possibility for resource management that is rationable from an economic and environmental perspective.

Arms, Karen. *Environmental Science*. Fort Worth, TX: Saunders, 1994. Provides a concise treatment of the main environmental problems and discusses their solutions.

Bramwell, Anna. *Ecology in the 20th Century: A History*. New Haven, CT: Yale University Press, 1989. Contains insightful examination of the political and spiritual origins of the ideas behind the growth of the environmental movement from 1880 to the 1980s. It proposes that today's environmental movement emerged from a politically radicalized environmentalism, predicated on the shift from mechanistic to vitalistic philosophy in the late nineteenth century.

Brown, Lester, R., et al. *State of the World*. New York: Norton, 1998; yearly. Provides an annual environmental picture of the world. The distinguished global environmental research team of the Worldwatch Institute analyzes the environmental conditions and progress toward sustainable society. It contains the latest data on an array of environmental issues.

Cable, Sherry, and Charles Cable. *Environmental Problems/Grassroots Solutions*. New York: St. Martin's Press, 1995. Offers readers an analysis of the sociological dimensions of environmental issues. It evaluates the roles that environmentalism has played in fighting pollution.

Caldwell, Lynton K. *International Environmental Policy*. Durham, NC: Duke University Press, 1991. Is an examination of the global diplomacy, treaties, and conventions for environmental protection.

Carson, Rachel. *Silent Spring*. Boston: Houghton Mifflin, 1962. This classic book alerted the public to the dangers of pesticide misuse and began a new wave in environmental activism in the 1960s.

Chisholm, Anne. *Philosophers of the Earth: Conversations with Ecologists*. New York: E. P. Dutton, 1972. Compiles information and personal interviews with key environmentalists of the twentieth century.

Commoner, Barry. *The Closing Circle*. New York: Alfred A. Knopf, 1971. A classic that, during the early stages of the antipollution concerns, warned the general public of the dangerous effects of the misuse of technology.

Commoner, Barry. *Making Peace with the Planet*. New York: New Press, 1992. Offers a comprehensive analysis of the role that technology plays in culture and environment. It prescribes what should be done to turn things around and to control technology's impact.

Council on Environmental Quality. *Annual Report of the Council on Environmental Quality*. Washington, DC: U.S. Government Printing Office, 1998; yearly. Is an annual compilation of data and reports of the Council on Environmental Quality as mandated by the National Environmental Policy Act of 1969. Each publication is transmitted to the U.S. Congress, and it evaluates current and foreseeable trends in environmental quality and the effects of these trends on the social, economic, and other requirements of the United States.

Cunningham, William P. *Understanding Our Environment*. Dubuque, IA: Wm. C. Brown, 1994. Introduces the central concepts of environmental science that will help readers understand current environmental issues.

Daly, Herman E., ed. *Steady-State Economics*. Washington, DC: Island Press, 1991. Has collection of essays on the physical and ethical dimensions of the steady-state economy. Contributors include economists, ecologists, environmental scientists, and others.

Dowie, Mark. *Losing Ground: American Environmentalism at the Close of the Twentieth Century*. Cambridge, MA: MIT Press, 1995. Describes the successes and failures of American environmentalism from its conservationist origins to its antipollution stance of the 1960s and 1970s.

Eckersley, Robyn. *Environmentalism and Political Theory: Toward an Ecocentric Approach*. Albany, NY: SUNY Press, 1992. Delivers an informative account of environmental political theory.

Ehrlich, Paul R. *The Population Bomb*. New York: Ballantine Books, 1968. The tip of the iceberg in the modern concern about overpopulation. A prophetic warning of the hidden dangers of the "population explosion."

Ehrlich, Paul R., and Anne Ehrlich. *The Population Explosion*. New York: Simon & Schuster, 1990. The updated version of *The Population Bomb* with new statistics and scenarios on the relationship between population and resource.

Elliott, Jennifer A. *An Introduction to Sustainable Development*. London: Routledge, 1994. Reviews the aims, successes, and shortcomings of sustainable economic policies in the developing nations.

Goldfarb, Theodore D. *Taking Sides: Clashing Views on Controversial Environmental Issues*. Guilford, CT: Duskin, 1993. Is a debate-style volume that focuses on the pros and cons of different viewpoints with regard to the fate of nature and environmental quality.

Gore, Albert. *Earth in the Balance: Ecology and the Human Spirit*. Boston: Houghton Mifflin, 1992. Offers an interesting description of the modern environmental crisis as well as how we can best overcome it.

Hardin, Garrett. *Living Within Limits: Ecology, Economics, and Population Taboos*. New York: Oxford University Press, 1993. Continues with Hardin's insightful and comprehensive analysis of the current population crisis.

Hays, Samuel P. *Beauty, Health, and Permanence: Environmental Politics in the United States, 1955–1985*. New York: Cambridge University Press, 1987. Presents a historical analysis of socioeconomic and political dimensions of environmental issues.

Jenseth, Richard, and Edward E. Lotto, eds. *Constructing Nature: Readings from the American Experience*. Upper Saddle River, NJ: Prentice-Hall, 1996. Is a collection of essays by nature writers on how nature and society relate to each other.

Johnson, Stanley P. *World Population—Turning the Tide*. Norwell, MA: Kluwer Academic Publishers, 1994. Summarizes the development, problems, and prospects of family planning policies of different governments as well as the subsequent evolution of international organizations that sought to support family planning efforts.

Kamieniecki, Sheldon, Robert O'Brien, and Michael Clarke, eds., *Controversies in Environmental Policy*. Albany, NY: SUNY Press, 1985. Describes the interactions of politics, technology, and economics in environmental decision making.

Kaufman, Donald G., and Cecilia M. Franz. *Biosphere 2000: Protecting Our Global Environment*. New York: HarperCollins, 1993. Covers environmental concerns, concentrating on the societal and scientific dimensions necessary to understand the environmental crisis.

Kline, Benjamin. *First Along the River: A Brief History of the U.S. Environmental Movement*. San Francisco, CA: Acadia Books, 1997. Is an interesting narrative of the history of the United States environmental movement from the colonial period to the 1990s.

Lacey, Michael J., ed. *Government and Environmental Politics: Essays on Historical Development since World War Two*. Baltimore, MD: Johns Hopkins University Press, 1991. Describes the evolution of the new environmental values that sparked the environmental movement and looks at government's response to the changing social values.

Leopold, Aldo. *A Sand County Almanac*. New York: Oxford University Press, 1949. This environmental classic consists of a collection of essays on nature and conservation. It chronicles Leopold's thoughts, in almost poetic prose, about the relationship between humans and wildlife.

Lester, James P., ed. *Environmental Politics and Policy: Theories and Evidence*. Durham, NC: Duke University Press, 1989. Presents an analysis of the politics behind environmental decision making concentrating on the role of socioeconomic-political factors.

Malthus, Robert. *An Essay on the Principle of Population*. New York: Oxford University Press, 1994 edition. Now a classic on the relationship between food supply and population, this is the first book that popularized the dilemma created by explosive human population growth.

Mannion, Antoinette M., and Sophie R. Bowlby, eds. *Environmental Issues in the 1990s*. West Sussex, England: John Wiley & Sons, 1992. Describes the most important concerns regarding the interaction between human society and the environment in the 1990s.

Marsh, George P. *Man and Nature: On Physical Geography as Modified by Human Action*. New York: Harvard University Press, 1965 edition. From a historical perspective, it is an environmental classic. It alerted the public to the follies of overexploiting nature, and it set the stage for development of environmental values during the nineteenth century.

Mathews, Jessica T., ed. *Preserving the Global Environment: The Challenge of Shared Leadership*. New York: W. W. Norton, 1991. Describes environmental concerns, economics, and policies from the national and international viewpoints.

McIntosh, Robert P. *The Background of Ecology: Concept and Theory*. Cambridge: Cambridge University Press, 1985. An overview of the origins, development, and current problems of ecological concepts and theories.

McKinney, Michael L., and Robert M. Schoch. *Environmental Science: Systems and Solution*. Sudbury, MA: Jones and Bartlet, 1996. Provides an environmental primer on resource management and pollution control with data from around the world.

Meadows, Donnella H., et al. *Beyond Limits: Confronting Global Collapse and Envisioning a Sustainable Future*. Post Mills, VT: Chelsea Green Publishers, 1992. A sequel to *The Limits of Growth*, this is a more modern computer-model study of the interaction of population, resources, and pollution. It suggests that society can achieve ecological balance if it develops alternative sustainable lifestyles.

Meadows, Donnella H., et al. *The Limits to Growth*. New York: Universe Books, 1972. Is a classic work that used computer models to alert the public to limits of resources.

Millbrath, Lester W. *Environmentalists: Vanguard for a New Society*. Albany: SUNY Press, 1992. Describes the composition of the modern-day environmental movement, concentrating on its tactics and the role it plays in building a sustainable society.

Miller, G. Tyler, Jr. *Living in the Environment*. Belmont, CA: Wadsworth, 1997. Provides the reader with basic information on how various components of the Earth are interconnected and interdependent, while concentrating on national and international environmental problems.

Mungall, Constance, and Digby J. Mclaren, eds. *Planet Under Stress: The Challenge of Global Change*. Toronto: Oxford University Press, 1990. Is a comprehensive analysis of key global concerns, such as climate change and other environmental problems.

Muschett, F. Douglas, ed. *Principles of Sustainable Development*. Delray Beach, FL: St. Lucie Press, 1996. Provides a history and background of sustainable development, emphasizing the ecological, social, and economic dimensions of sustainability.

Nash, Roderick F. *The Rights of Nature: A History of Environmental Ethics*. Madison: University of Wisconsin Press, 1989. Describes the intellectual history of environmental ethics, in particular the notion that morality should also incorporate the relationship between nature and society.

Nebel, Bernard J., and Richard T. Wright. *Environmental Science*. Englewood Cliffs, NJ: Prentice-Hall, 1993. Covers sustainability and the current practices resulting in nonsustainable environmental impacts. Suggests how society can achieve sustainability.

Norton, Bryan G. *Toward Unity Among Environmentalists*. Oxford: Oxford University Press, 1991. Is an overview of the evolution of environmentalism and environmental policies in the United States up to the 1980s.

Odum, Eugene P. *Ecology and Our Endangered Life-Support Systems*. Sunderland, MA: Sinauer, 1989. Contains a good introduction to ecology and human concerns. The author uses ecological concepts as a unifying theme to study environmental problems.

Owen, Oliver S., and Daniel D. Chiras. *Natural Resources Conservation: Management for a Sustainable Future*. Englewood Cliffs, NJ: Prentice-Hall, 1995. Provides a background of the past, present, and future of resource conservation. It describes the societal changes required to attain a sustainable future.

Park, Chris, ed. *Environmental Policies: An International Review*. Dover, NH: Croom Helm, 1986. Analyzes the environmental concerns and assesses the policies formulated to solve them in some developed and developing nations.

Petulla, Joseph M. *American Environmentalism*. College Station: Texas A & M University Press, 1980. An interpretive history of the environmental revolution to the late 1970s, it particularly deals with the problems and prospects of the environmental tradition.

Pirages, Dennis C., ed. *Building Sustainable Societies*. Armonk, NY: M. E. Sharpe, 1996. This collection of essays evaluates whether the industrial model can achieve sustainability. It assesses the sociological and political ramifications of spiraling resource consumption.

Population Reference Bureau. *World Population Data Sheet and United States Population Data Sheet, 1998*; yearly. These are annual collections of demographic data of the United States and the world. They provide birth rates, death rates, and other relevant demographic data.

Raven, Peter H., Linda R. Berg, and George B. Johnson. *Environment*. Fort Worth, TX: Saunders, 1993. Offers a comprehensive analysis on the way nature works, and what is happening to the environment as human population mushrooms.

Rogers, Adam. *The Earth Summit: A Planetary Reckoning*. Los Angeles, CA: Global View Press, 1993. A contribution to our understanding of the factors that led to the Earth Summit. It selects a number of people who contributed to the Earth Summit, to explain, in their own way, what transpired during this international conference.

Rubin, Charles T. *The Green Crusade: Rethinking the Roots of Environmentalism*. New York: Macmillan, 1994. Covers the environmental movement from the 1960s to the 1990s, its causes and campaigns, its key proponents and enemies, its glories and failures.

Sale, Kirkpatrick. *The Green Revolution: The American Environmental Movement, 1962–1992*. New York: Hill and Wang, 1993. Examines the most influential environmental popularizers of the environmental movement from 1962 to 1992 to determine how their efforts were received and those who opposed and criticized the movement.

Santos, Miguel A. *Managing Planet Earth: Perspectives on Population, Ecology, and the Law*. Westport, CT: Greenwood Publishing Group, Inc., 1990. Analyzes the scientific and legal dimensions of environmental stability. It suggests a new world order based on ecological balance, which is politically implemented through an international environmental authority—a proposed seventh organ of the United Nations.

Scheffer, Victor B. *The Shaping of Environmentalism in America*. Seattle: University of Washington Press, 1991. Describes the history of the environmental movement, primarily the decades of the 1960s and 1970s, tracing the roots of the environmental movement and its relationship to other events, such as the social upheaval of the 1960s that was taking place at the time.

Schnaiberg, Allan, and Kenneth A. Gould. *Environment and Society: The Enduring Conflict*. New York: St. Martin's Press, 1994. Describes the powerful societal forces that impeded environmentally benign behavior and the ability of the environmental movement to overcome these societal forces. It concludes that a benign society-nature interaction is possible, if effective action is taken now.

Silver, Chery S., and Ruth S. DeFries. *One Earth, One Future: Our Changing Global Environment*. Washington, DC: National Academy of Science, 1990. Outlines some of the international implications of transboundary pollutants, such as global climate change, ozone depletion, and acid rain.

Southwick, Charles H., ed. *Global Ecology in Human Perspective*. New York: Oxford University Press, 1996. Examines the global dimensions of ecological concerns, particularly how society impacts the global commons and how these changes affect our health, behavior, and socioeconomic-political system.

Switzer, Jacqueline V. *Environmental Politics: Domestic and Global Dimensions*. New York: St. Martin's Press, 1994. A synthesis and critique of national and international environmental policies.

United States Bureau of the Census. *Current Population Reports*. Washington, DC: U.S. Government Printing Office, 1998; yearly. Contains statistics on birth, immigration, and changing age structure of the United States.

United States Bureau of the Census. *Statistical Abstract of the United States*. Washington, DC: U.S. Government Printing Office, 1998; yearly. Contains charts that provide statistics on population, health, production, agriculture, trade, foreign aid, and other pertinent information.

Vig, Norman J., and Michael E. Kraft, eds. *Environmental Policy in the 1990s*. Washington, DC: Congressional Quarterly, 1994. Describes the key concerns of environmental policy and politics, primarily from 1960 to 1990, and evaluates the underlying controversies that distort environmental policy-making.

White, Rodney R. *North, South, and the Environmental Crisis*. Toronto: University of Toronto Press, 1993. Contains an overview of the basic scientific challenges taking place in the environment comparing the way the relationship between rich and poor countries will be affected by the internationalization of environmental issues.

Wilson, Edward O. *The Diversity of Life*. Cambridge: Harvard University Press, 1992. Is a comprehensive analysis of the crisis caused by the extinction of species in the twentieth century.

World Commission on Environment and Development. *Our Common Future*. Oxford: Oxford University Press, 1987. A landmark publication that first brought worldwide attention to the concept of sustainable development. The report covers the political and economic problems of achieving sustainable development and identifies international strategies for managing the global commons.

World Resources Institute. *World Resources 1998*. New York: Oxford University Press, 1998; yearly. This annual report, by the World Resources Institute in collaboration with the United Nations Environment Programme, contains about fourteen chapters and data on a wide range of environmental issues.

VIDEOS

Davis, David. *Out of the Ozone*. (19 minutes.) Huntsville, TX: Educational Video Network, 1991. Describes the development of the ozone shield and its importance to the evolution of life on Earth.

Davis, David. *World Environmental Cooperation*. (19 minutes.) Huntsville, TX: Educational Video Network, 1994. Starts by briefly looking at the formation of the Earth's biosphere and ends by providing a comprehensive description of the problems of managing the global commons, such as global commons and Antarctica.

Davis, David. *World Population Problems*. (17 minutes.) Huntsville, TX: Educational Video Network, 1994. A description of the root of the population crisis, concentrating on the ecological characteristics of human population. It deals with the concept of carrying capacity, age structure analysis of developed and developing nations, and India's and China's population policies.

Hurst, Sam (producer). *Paul Ehrlich and the Population Bomb*. (55 minutes.) San Francisco, CA: KQED, 1996. Interesting video on the impact that Paul Ehrlich's visit had on the development of his idea that the primary cause of pollution is overpopulation. Looks at the pros and cons of the environmental movements that emerged during the 1960s. Lively expression of ideas by Judith Bruce, Rachel Carson, Barry Commoner, Paul and Anne Ehrlich, John Holdren, Peter Raven, and Julian Simon. Hosted by David Suzuki, who provides insightful commentary and analysis.

International Video Publications. *Save the Earth: A How-To Video*. (60 minutes.) Huntsville, TX: Educational Video Network, 1990. A litany of things that individuals can do to save the environment as well as a description of environmental concerns. Published with the association of the Save the Earth Brigade. Introduced by the former Prime Minister of England Margaret Thatcher and narrated by Jere Burns.

Nolan, Mary L. *Saving Forest Ecosystems: World Forests in Danger.* (34 minutes.) Huntsville, TX: Educational Video Network, 1995. Brief description of different forest ecosystems and the problems of deforestation. Includes a discussion of the natural selection logging and the importance of the forest to human society.

Smithsonian Institution. *Our Biosphere: The Earth in Our Hands.* (45 minutes.) Huntsville, TX: Educational Video Network, 1991. A detailed and interesting look at the Biosphere II project, which was originally conceived as an experiment for building a prototype for human planetary missions. The video describes the problems and prospects of building a microcosm of the Earth's ecosphere. Narrated by Robert Redford.

World Resources Institute. *Introduction to Our Global Environment.* (10 minutes.) Washington, DC: World Resources Institute, 1994. A brief environmental education project concentrating on the links between environmental and social problems, e.g., energy use and global warming.

World Resources Institute. *Preserving Our Global Environment.* (53 minutes.) Washington, DC: Schecter Films, 1994. An interchange of ideas by leaders in government, labor, and the environment on the interrelationships between population growth, biodiversity, and global climate change. Hosted by Jessica Tuchman and David Gergen.

INTERNET SOURCES

AltaVista Search Engine: http://altavista.digital.com/—General source of environmental information.

Cornell University Law School: http://www.law.cornell.edu/topics/environmental.html—Law topics bearing on federal agencies with environmental responsibilities, Supreme Court decisions, and environmental statutes.

Council on Environmental Quality NEPA Net: http://ceq.eh.doe.gov/nepa/nepanet.htm—Full text of the National Environmental Policy Act (NEPA), regulation for implementing NEPA, agency Web sites, and guidance from the Council on Environmental Quality.

Environet: http://www.aspenlinx.com/environet/—Links to environmental information.

Environmental Gophers: gopher.//path.net:8001/11/.subject/Environment/—General source of environmental information.

Exploring Internet: http://www.globalcenter.net/gcweb/tour.html—General source of environmental information.

Global Resource Information Database: http:www.grid.unep.ch/grid-home.html/—General source of environmental information.

Governmental World Wide Web Resources: http://www.environlink.org/environ-gov.html/—General source of environmental information.

Information Center for the Environment: http://ice.ucdavis.edu/—A cooperative effort of environmental scientists at the University of California, Davis, and collaborators at private, state, federal, and international organizations interested in environmental protection.

Infoterra: United Nations Environment Programme (UNEP): http://pan.ce-dar.univie.ac.at/gopher/unep/—Main structure: databases and resources, UNEP, and environmental links.

Institute for Global Communications: http://www.igc.org/igc/—Projects include environmental sustainability.

National Library for the Environment: http://www.cnie.org/nle/—On-line components include congressional research service reports, in-depth issues in the environment, and population and environmental linkages.

Natural Resources Defense Council: On Line: http://www.nrdc.org/html—Includes news and information, online journal, technical materials, and eco guide.

Natural Resources Research Information Pages: http://sfbox.vt.edu:10021/Y/yfleung/nrips.html/—General information about natural resources.

Population Reference Bureau: http://www.prb.org/prb/—Information on a wide range of topics, including U.S. and international population trends.

United Nations Development Programme (UNDP): http://www.undp.org/—UNDP text version Web site. Focus areas include information about sustainable human development.

United Nations Population Information Network (POPIN): http://www.undp.org/popin/popin.htm —Contains information on world population trends and International Conference on Population and Development.

Usenet Keyword Search: http://www.excite.com/—General source of environmental information.

World Wide Web Virtual Library: Environment: http://earthsystems.org/Environment.shtml—Subjects include biodiversity and ecology, energy, environmental law, and sustainable development.

Yahoo! Search Engine: http://www.yahoo.com/—General source of environmental information.

Index

About the Author

MIGUEL A. SANTOS is Professor of Ecology and Environmental Science at Baruch College, New York, where he is the coordinator of environmental studies. Since 1975 he has taught a wide variety of courses, focusing especially on ecology, environmental science, and law. He is the author of *Managing Planet Earth* (Bergin & Garvey, 1990) and six other books, plus many publications in scientific and legal journals.